Chemical Bonds: A Dialog

Inorganic Chemistry
A Textbook Series

Jeremy Burdett
Curriculum vitae

Born in London in 1947. After receiving a BA from Cambridge University in 1968 he spent two years at the University of Michigan as Power Foundation Fellow. In 1970 he returned to Cambridge as an SRC Postdoctoral Fellow and received his PhD in 1972. In the same year he was appointed Senior Research Officer at The University of Newcastle upon Tyne. During these years he worked on aspects of the electronic structure and properties of small molecules, with special focus on the correlation between electronic and geometrical structure. After a sabbatical leave at Cornell University with Roald Hoffmann he moved to The University of Chicago in 1978. He is presently Professor and Chairman in the Department of Chemistry and has a joint appointment with The James Franck Institute. His interests since arriving at Chicago have been largely theoretical in nature and have been centered mainly around the structures of solids. Present interests include: high-temperature superconductors and metal–insulator transitions in general, disordered structures and non-stoichiometry, and the design of materials with specific properties. He has been a Sloan Fellow, a Camille and Henry Dreyfus Teacher-Scholar, a Fellow of The John Simon Guggenheim Memorial Foundation, was awarded The Meldola Medal and Prize of The Royal Institute of Chemistry in 1977 and in 1993 the Amoco Foundation Award for Distinguished Contributions to Undergraduate Teaching. He was the Wilsmore Fellow at The University of Melbourne in 1985 and was CNRS Visiting Professor, Université de Paris-Sud, Orsay in 1987 and at Rennes in 1994. He was awarded the ScD degree by The University of Cambridge in 1991 and the Tilden Medal and Prize for 1995 by The Royal Society of Chemistry. He is the author of some 200 research publications and several books. The latter include: *Molecular Shapes: Theoretical Models of Inorganic Stereochemistry*, John Wiley & Sons, Inc. (1980), *Orbital Interactions in Chemistry* (with Thomas A. Albright and Myuan-Hung Whangbo), John Wiley and Sons, Inc. (1985), *Problems in Molecular Orbital Theory* (with Thomas A. Albright), Oxford University Press (1992) and *Chemical Bonding In Solids*, Oxford (1995).

Chemical Bonds: A Dialog

Jeremy Burdett

The University of Chicago

John Wiley & Sons
Chichester · New York · Weinheim · Brisbane · Singapore · Toronto

Other Wiley Editorial Offices

John Wiley & Sons, Inc., 605 Third Avenue,
New York, NY 10158-0012, USA

VCH Verlagsgesellschaft mbH, Pappelallee 3,
D-69469 Weinheim, Germany

Jacaranda Wiley Ltd, 33 Park Road, Milton,
Queensland 4064, Australia

John Wiley & Sons (Asia) Pte Ltd, 2 Clementi Loop #02-01,
Jin Xing Distripark, Singapore 129809

John Wiley & Sons (Canada) Ltd, 22 Worcester Road,
Rexdale, Ontario M9W 1L1, Canada

British Library Cataloguing in Publication Data

A catalogue record for this book is available from the British Library

ISBN 0 471 97129 4; 0 471 97130 8 (pbk.)

Typeset in 10/12pt Times by Techset Composition Ltd, Salisbury
Printed and bound in Great Britain by Bookcraft (Bath) Ltd
This book is printed on acid-free paper responsibly manufactured from sustainable forestation,
for which at least two trees are planted for each one used for paper production.

Contents

Prologue

There is a little book I came across the other day that I would like to tell you about. In 1948 William Hume-Rothery, later to become Professor of Metallurgy at The University of Oxford, published a book[1] that was set around a Socratic interlude between two scientists, 'the Old Metallurgist' and 'the Young Scientist'. Over a period of years Hume-Rothery had published several books that strove to educate the metallurgical community in the new quantum mechanical methods that were then just beginning to have a striking impact on the field. The idea of a question and answer tutorial, where the Young Scientist leads the Old Metallurgist step by step through the quantum mechanical description of atoms, then molecules and eventually to metals, is a very useful one and the book is well-worth reading even today. In a later edition (1963) 'the Old Metallurgist' has become 'the Retired Metallurgist' and 'the Young Scientist', 'the Middle-Aged Scientist'. The two reminisce (in a series of 'Epilogues') over the progress since the first edition. Although there are few similarities between the present status of the theory of Chemical Bonding and that of Physical Metallurgy when this book was written, perhaps it's a useful format for discussing aspects of molecular structure.

Yes, the status of the theory of chemical bonding in molecules and solids today is very different from that of the metallurgical field in 1948. We presently have quite a good understanding of many aspects of the structures of molecules and solids. Hume-Rothery was clearly trying to encourage materials scientists to advance their field by going beyond their somewhat arbitrary foundry practices to thinking about some of the physical reasons behind the structures and properties of materials.

There are some similarities, though. Many of the theoretical concepts that we use in describing molecules are used in a somewhat arbitrary way just like the Old Metallurgist in the foundry. Also, results that come today from the computer very often have never been processed by any grey matter at all. Surely there is much to be gained by seeing the origin of even the most fundamental results. Charles Coulson, a contemporary of Hume-Rothery at Oxford, wrote; 'The rôle of quantum chemistry is to understand the elementary concepts of chemistry and to show what are the essential features of chemical behaviour'[2]. The important word here is surely 'understand'.

Yes, you are quite right. Understanding the origin of theoretical results in any area is important, but being able to weave a linking theoretical thread between diverse areas is particularly so. Thus when looking at chemical bonding problems, it is rewarding to be able to use the same model throughout. In particular the orbital model has proven very effective in many areas over the years. But remember that the origin of many chemical results, especially those involving bond energies, reactivity and whether a given solid is a metal or an insulator, for example, lies buried in extremely complex calculations and even if we can do the calculation itself, a chemical picture is often difficult to extract. As Robert Mulliken wrote; 'the more accurate the calculations become, the more the concepts tend to vanish into thin air'[3].

Well yes, I know that not all of chemistry's problems are soluble using existing theoretical ideas. In fact, early on in Hume-Rothery's book, The Young Scientist admits that his quantum mechanical ideas are not immediately going to lead to ways to make stronger alloys and thus lead to a product of tangible value to the Old Metallurgist.

Perhaps then we should take some aspects of chemical bonding, each of importance to chemists and study them in the same way Hume-Rothery's characters did. There are many fundamental topics we shall not need to study, since I think you have a pretty good understanding of them already. I don't think you are quite as out of date as the Old Metallurgist was. We will assume a basic knowledge of the electronic structure of atoms and molecules but will present the mathematical structure of molecular orbital theory in Chapter 2. The discussion of some of the material will assume a basic knowledge of perturbation theory and how it is used to understand molecular structure. It is a little more difficult to grasp.

So we should therefore be aware that, while some of the topics we will discuss are relatively easy to appreciate, some are much more complex. However, as the Young Scientist said 'It is not so much a question of difficulty or time, as of being prepared to think. The subject is not one to read in spare time in railway carriages, or while lunching in a restaurant.'

References

1. Hume-Rothery, W., *Electrons*, *Atoms*, *Metals and Alloys*, Published for *Metal Industry* by The Louis Cassier Co. Ltd., Distributed by Iliffe and Sons Ltd., and by The Philosophical Library, New York (1948, 1963).
2. Coulson, C. A., *Rev. Mod. Phys.*, **32**, 190 (1960).
3. Mulliken, R. S., *J. Chem. Phys.*, **43**, S2 (1965).

1 What is the Nature of the Chemical Bond?

Like most chemists I am sure that I take for granted the very existence of chemical bonds in molecules and solids. In fact I use the concept in almost everything I do, thinking about structure, reactivity and properties. But what are the physical underpinnings behind their formation? This must be the most fundamental aspect of chemistry. Let's start with the most popular picture. Molecular orbital theory draws out the level diagram for H_2 and H_2^+ as in Figure 1.1. A bonding orbital is generated lying deeper in energy than its antibonding partner and deeper than the energies of the orbitals in the free atom. Simple Hückel theory writes the stabilization energy of the bonding orbital as $\beta = \langle \phi_1 | \mathscr{H}^{eff} | \phi_2 \rangle$ or $\int \phi_1 \mathscr{H}^{eff} \phi_2 d\tau$ where $\phi_{1,2}$ are the two hydrogenic 1s basis functions. (Let's use the former notation in our discussions.) $\mathscr{H}^{eff}$ is some effective Hamiltonian that we will see more of in Chapter 2. Although it's clear that $\langle \phi_1 | \mathscr{H}^{eff} | \phi_2 \rangle$ leads to a stabilization of the system, does the elegance of quantum mechanics conceal the details of the bond-forming process? What, for example, is the interplay between the potential and kinetic energy that gives rise to this?

The question of what determines the binding energy of molecules has a long history. Unfortunately, many of the explanations that have been given are not right. In fact the answer is relatively straightforward, but does involve some considerations that are not immediately obvious. The classic work on this fundamental topic is a paper by Klaus Ruedenberg[1] that has been nicely interpreted for the layman by Colin Baird[2]. Let's start though with perhaps the simplest model (which we can easily explore algebraically) by first imagining the energy levels of the hydrogen atom as described by those of a particle in a cubical box of side a, and then imagine formation of the H_2^+ molecule by bringing together two such boxes so that the dimensions of the box are $a \times a \times 2a$ as in Structure **1.1**. Recall[3] that the energy levels for a particle in a box of length $a \times b \times c$ are labeled by three quantum numbers, (n_1, n_2, n_3). The level energies are given by

$$E(n_1, n_2, n_3) = \frac{h^2}{8m}\left(\frac{n_1^2}{a^2} + \frac{n_2^2}{b^2} + \frac{n_3^2}{c^2}\right) \tag{1.1}$$

where m is the electron mass. The normalized wavefunctions are readily evaluated as

$$\psi(n_1, n_2, n_3) = \sqrt{\frac{8}{abc}} \sin\left(\frac{n_1 \pi x}{a}\right) \sin\left(\frac{n_2 \pi y}{b}\right) \sin\left(\frac{n_3 \pi z}{c}\right) \tag{1.2}$$

where the $n_i = 1, 2 \ldots$. For the cubical box the lowest energy level has an energy of $3X$ where $X = h^2/8ma^2$. Its wavefunction along x is shown in Figure 1.2(a). For the

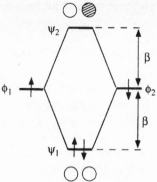

Figure 1.1
Molecular orbital diagram using the Hückel approximation for H_2

rectangular box, $a \times a \times 2a$, which we are going to use as a model for H_2^+, the energy is lower. You can show that it is just $2.25X$, i.e., a 'bond' has been formed. Now the 'box' we have just described is a special one in that the potential energy is fixed, except at the walls where it is infinite. (Inside the box we can arbitrarily set it to zero.) Thus it is the reduction in *kinetic* energy which leads to a stabilization of the doubled box.

In general the kinetic energy is given by

$$T(n_1, n_2, n_3) = \frac{\langle \psi | [-(\hbar^2/2m)\nabla^2] | \psi \rangle}{\langle \psi | \psi \rangle} \tag{1.3}$$

$$= \frac{\langle (\hbar^2/2m)[\nabla\psi]^2 \rangle}{\langle \psi | \psi \rangle} \tag{1.4}$$

Calculation of T using the wavefunction of equation (1.2) leads to equation (1.1) of course, but there is a nice pictorial aspect of this formulation. Notice from equation (1.3) that the kinetic energy is determined by the slope of the wavefunction. Observe that in Figure 1.2(b) the average slope has decreased on making the box larger. One can imagine an infinite box with a zero slope and thus zero kinetic energy.

The simplest view of molecular orbital theory leads to some related results. The molecular orbital model writes the orbitals of a molecule (ψ_i) as a simple linear combination of the atomic orbitals (LCAO), ϕ_j, of the atoms of which it is composed.

$$\psi_i = \sum_j c_{ij} \phi_{ij} \tag{1.5}$$

I used this approximation in Figure 1.1, but what is the justification for the approach?

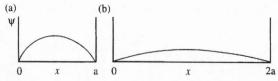

Figure 1.2
Particle-in-a-box wavefunction along x. In (a) the box is of side a and smaller than in (b), 2a

It is indeed an approximation. A more accurate state of affairs would not impose such a limitation in terms of the number and type of function in this sum. But a working rationale for using this simple approach lies in the fact that the electron density in the molecule (as determined experimentally by accurate X-ray diffraction studies, for example) shows only a small deviation from the electron density of the constituent atoms. Much of the essence of chemical bonding is captured by the approximation too[4]. Indeed in quite sophisticated numerical models of molecules and solids, the sum of the atomic densities is used as a starting point for a more accurate calculation.

Using this prescription for the diatomic H_2^+, its molecular orbitals may be written as

$$\psi = a\phi_1 + b\phi_2 \tag{1.6}$$

where $\phi_{1,2}$ are the two atomic hydrogen 1s orbitals. The electron density function which describes equation (1.6) is then

$$\psi^2 = a^2\phi_1{}^2 + b^2\phi_2{}^2 + 2ab\phi_1\phi_2 \tag{1.7}$$

The ϕ_i are real so we don't need to use the complex conjugate here, i.e., $\psi^2 = \psi^*\psi$. Since by symmetry the electron density must be equally shared by both hydrogen atoms, $a = \pm b$. This is a very simple but powerful result. In more complex molecules the form of the wavefunction is determined by similar considerations approached using the techniques of Group Theory. Writing $S = \langle \phi_1 \mid \phi_2 \rangle$, as the overlap integral between ϕ_1, ϕ_2, normalized bonding, ψ_b and antibonding, ψ_a, orbitals result

$$\psi_b = 1/\sqrt{2(1+S)}(\phi_1 + \phi_2)$$
$$\psi_a = 1/\sqrt{2(1-S)}(\phi_1 - \phi_2) \tag{1.8}$$

These are shown in Figure 1.3. Notice that the average slope of the wavefunction describing the bonding orbital is less than that for the antibonding orbital. Focus especially on the magnitude of the first derivative of the wavefunction at the point half-way between the nuclei. Certainly on this basis the bonding orbital is more stable than the antibonding orbital on kinetic energy grounds. In fact, since there is less charge between the nuclei in

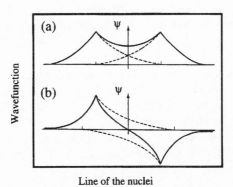

Line of the nuclei

Figure 1.3
Bonding, ψ_b and antibonding, ψ_a, orbitals for H_2

the antibonding orbital than in its bonding counterpart, one could say that their relative stability was also set by a poorer potential energy term in ψ_a.

It looks then as though the chemical bond comes about through the availability of a larger 'box' that the electrons occupy in the bonded state, leading to a reduction in kinetic energy.

Unfortunately this simple model is not right. The picture comparing the *relative* energies of bonding and antibonding orbitals is a correct one but this simple view of the chemical bond itself as formed from particles in boxes, is just too simple, although it can be modified to give the correct result[5]. Just as in many parts of chemistry, things are not quite what they seem. We will see that it is the change in potential energy of the electrons that leads to bonding, not the kinetic energy as suggested by this simple model. That this is true is indicated in a completely general way by the Virial Theorem.* This fundamental concept relates the kinetic (T) and potential (V) energy (under special circumstances as we will see later) as $V = -2T$. The total energy is the sum $E = V + T$.

So for the molecule

$$V_{\text{mol}} = -2T_{\text{mol}} \tag{1.9}$$

and for the atoms

$$V_{\text{at}} = -2T_{\text{at}} \tag{1.10}$$

Since the change in total energy on molecule formation is just

$$\Delta E = \Delta T + \Delta V \tag{1.11}$$

then

$$\Delta E = -\Delta T = +\Delta V/2 \tag{1.12}$$

Since ΔE must be negative for formation of a stable molecule, then ΔT must be positive and ΔV negative; i.e., it's the change in the potential energy that is behind the energetics of bond formation.

That's quite right. However, to understand how this comes about in a real molecule[2] it's useful to look at the simplest species first, namely the atom. We start by looking at the potential energy associated with the attraction of the electron to the nucleus of charge $+Z$. (We will work in atomic units (au) by putting the electronic charge and electron mass equal to unity.) It is simply given by Coulomb's law as $V = -Z/r$, where r is the electron–nucleus distance. The average potential energy is thus $V_{\text{av}} = -Z(1/r)_{\text{av}} = -Z\chi$. Thus the closer the electron gets to the nucleus, the larger (more negative) its potential energy. We saw earlier how the kinetic energy of the electron depends inversely upon the square of the size of the box in which it is contained. The same is true for the spherical hydrogen atom. With $Z = 1$ the average kinetic energy is $T_{\text{av}} = \frac{1}{2}(1/r)_{\text{av}}^2 = \frac{1}{2}\chi^2$. The potential energy is $V_{\text{av}} = -(1/r)_{\text{av}} = -\chi$. Using this approach we can readily visualize a simple model for the dynamics of the atom. Although the potential energy leads to rapid stabilization with decreasing r, the kinetic energy increases too. The average distance at equilibrium will be set by the balance of the two. You can differentiate the total energy ($E = -\chi + \frac{1}{2}\chi^2$) with respect to r, set it equal to zero to get this. You will find that $\chi = (1/r)_{\text{av}} = 1$. Thus the equilibrium value of r_{av} is one Bohr unit or one atomic unit (au). Figure 1.4 shows the

* The virial theorem for classical interaction is:

For COULOMB (INTERACTION) $V = -2K$ b 8π

S.H.M. $V = K$. Also true for Quantum interaction

(see H.C. B/P K, T68)

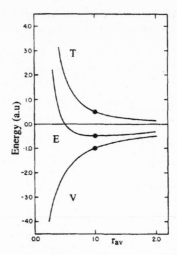

Figure 1.4
Behavior of the electronic potential and kinetic energy terms with respect to r_{av}, the average electron–nucleus distance, for the hydrogen atom. The solid circles represent the lowest (total) energy equilibrium situation. It is at this point only that the Virial Theorem is obeyed

behavior of the two terms with respect to r. The Virial Theorem is obeyed $(V = -2T)$ but only after having minimized the energy in this way when $\chi = (1/r)_{av} = 1$. When $\chi = (1/r)_{av} \neq 1$ the theorem is not obeyed.

A similar situation arises with earth (or other) bound satellites. The radius of their orbit is determined by the balance of kinetic and potential energy (in this case gravitational) in the same way. The Virial Theorem is obeyed for this situation, but only for this stable orbit. Clearly the theorem is not obeyed for some arbitrary motion, such as that of a speeding meteorite on a collision course with the earth. This is the special condition we mentioned earlier concerning the use of the theorem.

One can see that if for some reason in the atom the kinetic energy were reduced, then $\chi = (1/r)_{av}$ would increase, the average distance between electron and nucleus would decrease, and the potential energy would increase in magnitude. This point will be very important later.

If we now move on to the molecule H_2^+, in terms of potential energy changes there are two new terms to consider. There are the effects of nuclear–nuclear repulsion, $1/R$ (where R is the internuclear separation) and of the attraction of the electron for both nuclei.

Roughly speaking, the extra attraction and the repulsion cancel each other. The kinetic energy, however, is reduced due to the larger box in which the electron can move. This is just the situation about which we hypothesized above. The response of the system to the lower kinetic energy is to reduce the potential energy too. In fact the potential energy may be reduced (increased in absolute magnitude) quite a bit by shrinking the atoms (a larger χ or $(1/r)_{av}$). If the potential energy is reduced sufficiently, then it will become the dominating term in the energetics of bond formation. This is actually what happens. It sounds complex in words. Let's see how this comes about algebraically[2].

Recalling our discussion for the atom:

$$V = -\chi$$
$$T = \tfrac{1}{2}\chi^2$$
$$E = -\chi + \tfrac{1}{2}\chi^2$$
$$\partial E/\partial \chi = -1 + \chi = 0 \qquad \chi = 1 \tag{1.13}$$
$$V = -1.0$$
$$T = 0.5$$
$$E = -0.5 \text{ au}$$

For the molecule:

$$V \approx -\chi + 1/R - 1/R = -\chi$$
$$T = \alpha(\tfrac{1}{2}\chi^2) \qquad \alpha < 1 \tag{1.14}$$
$$E = -\chi + \alpha(\tfrac{1}{2}\chi^2) \qquad \chi = 1$$

If $\alpha = 5/6$ (since this is less than unity the particle is moving in a larger 'box'), $E = -0.583$ au or a binding energy of 0.583 au $- 0.5$ au $= 0.083$ au (220 kJ/mol). So keeping χ fixed at 1.0 au leads to a decrease in the kinetic energy and a stabilization of the system just as we would have expected from our earlier discussion. But has $\chi = (1/r)_{av}$ changed from its value in the free atom as a result of its new environment? Let's minimize the total energy with respect to χ to see:

$$V + T = E = -\chi + \alpha(\tfrac{1}{2}\chi^2)$$
$$\partial E/\partial \chi = -1 + \alpha\chi = 0 \qquad \chi = 1/\alpha = 6/5$$
$$V = -1.20 \text{ au}$$
$$T = 0.6 \text{ au} \tag{1.15}$$
$$E = -0.6 \text{ au or a binding energy of } 0.1 \text{ au (265 kJ/mol)}$$

This last result should be compared with the experimental value of 270 kJ/mol. This splendid agreement with experiment comes from our choice of α of course but there are more weighty reasons behind its selection that you can read about in references 1, 2. Notice by a comparison of equations (1.13) and (1.14) that the penalty associated with the increase in kinetic energy has been more than compensated by the stabilization associated with the change in the potential energy. Note too, that now the Virial Theorem is obeyed. This comes about since, by analogy with the satellite example, we have now identified the lowest energy, or stable motion, of the system. Thus this model of chemical bond formation shows that atoms are not static objects but contract during formation of the bond. Ideas that do not build in this relaxation will invariably show that it is the kinetic energy that is the important term.

But let's come back to the orbital model of Figure 1.1. How do these ideas fit in with this picture? It is one which will form the basis for much of our discussion in the rest of the book and so it's important to understand it in as much detail as possible.

There is in fact a direct extension of these ideas to the orbital model. The bonding orbital in H_2 or H_2^+ is given by the in-phase component of equation (1.8) using the LCAO philosophy of equation (1.5). Now, for the hydrogen atom $\phi_1 = N \exp(-r_A)$ where r_A is the electron-to-nucleus-A distance in atomic units and N is a normalization constant. To evaluate the energy associated with this wavefunction we need the Hamiltonian operator

$$\mathcal{H} = -\nabla^2/2 - 1/r_A - 1/r_B + 1/R \tag{1.16}$$

A little algebra leads to a value[6] for the total energy of H_2^+ as

$$E = E_H + \frac{1}{R} + \left\{ \frac{\left(1 + \frac{1}{R}\right)\exp(-2R) - (R+1)\exp(-R) - \frac{1}{R}}{1 + \left(1 + R + \frac{R^2}{3}\right)\exp(-R)} \right\} \tag{1.17}$$

where E_H is the energy of the free hydrogen atom. At infinite R this function has the correct form, $E \to E_H$. The minimum in the function of (1.17) with respect to R is found at 1.32 Å, far from the experimental distance of 1.06 Å. The error in the dissociation energy is about 37%. We can calculate[7] the kinetic and potential energies separately from equation (1.17). For the H_2^+ ion at its minimum along R, $E = -1.125$ au, $T = 0.763$ au and $V = -1.888$ au. For the separated species $E = -1.0$, $T = 1.0$ au and $V = -2.0$ au, a result which indicates that it is the reduction in kinetic energy which stabilizes the molecule relative to its constituents just as we found above when bringing together the two boxes of Structure **1.1**. However, note one very important aspect of this result. As with its analogs in the discussion earlier, it is in serious disagreement with the Virial Theorem (recall that T should equal $-E$) and therefore is one which cannot be trusted.

If a 'scaled' wavefunction for the hydrogen atom is used of the form $\phi_1 = N \exp(-\xi r_A)$ the energy may be minimized as a function of ξ. The Variational Theorem tells us that, although this energy will always be higher than the true energy, it will be the closest available for that choice of wavefunction. The use of such a scaled wavefunction and the theorem has another tremendous advantage which is not an obvious one. As shown by Vladimir Fock, the resulting energies obey[8] the Virial Theorem in a similar way to the introduction of a variational step in our earlier discussions. Minimization of the energy with respect to ξ in this way for the H_2^+ ion leads to a value of $\xi = 1.228$ and an energy minimum along R at 1.06 Å, a figure in exact agreement with experiment. The error in the dissociation energy is now reduced to 19%. At this point $E = -1.165$ au, $T = 1.165$ au and $V = -2.330$ au, a result in perfect agreement with the Virial Theorem. Note that it is now the increase in the magnitude of the potential energy which determines the stabilization of the H_2^+ ion just as we concluded earlier. We can also see that the origin of the result lies in the contraction ($\xi = 1.228$) of the hydrogen 1s orbitals relative to the free atoms ($\xi = 1.000$). So the details of the orbital picture fit in very neatly with our other discussions. So when considering this very important topic it is vital to have a quantum mechanical result in agreement with the Virial Theorem, otherwise the wrong inferences will be made. Given the form of the bonding molecular orbital one can indeed see that it is the concentration of charge between the nuclei demanded by ψ_b of equation (1.8) which leads to the chemical bond via a decrease in the potential energy of the system.

There are other views of the origin of chemical bonding in textbooks, presumably which are incorrect.

Yes, one common statement is that the chemical bond arises through the exchange energy. It was the foundation of Valence Bond Theory. You can read in reference 9 why it is incorrect.

Does the same model apply to all chemical bonds?

This is a good question. We presume so, but other than the restrictions of the Virial Theorem, we do not have a comparable analysis for these much more complex systems.

References

1. Ruedenberg, K., *Rev. Mod. Phys.*, **34**, 326 (1962).
2. Baird, N. C., *J. Chem. Educ.*, **63**, 660 (1986).
3. There are many places to find the origin of these. One is Atkins, P. A., *Quanta, A Handbook of Concepts*, Oxford University Press (1991).
4. Albright, T. A., Burdett, J. K. and Whangbo, M.-H., *Orbital Interactions in Chemistry*, John Wiley & Sons (1980).
5. Kuhn, H., Försterling, H.-D., *Principles of Physical Chemistry Understanding Molecules, Aggregates and Supramolecular Machines*, John Wiley & Sons (1997).
6. Hirschfelder, J. O. and Kincaid, J. F., *Phys. Rev.*, **52**, 658 (1937).
7. Linnett, J. W., *Wave Mechanics and Valency*, Methuen (1960).
8. Fock, V., *Zeits. Phys.*, **63**, 855 (1930).
9. McWeeney, R., *Spins in Chemistry*, Academic Press (1970).

General References and Further Reading

Atkins, P. A., *Molecular Quantum Mechanics*, Oxford University Press (1983).

Atkins, P. A., *Physical Chemistry*, Oxford University Press (1994).

Berry, R. S., Rice, S. A. and Ross, J., *Physical Chemistry*, John Wiley & Sons (1963).

DeKock, R. L., *J. Chem. Educ.*, **64**, 934 (1987).

Feinberg, M. J., Ruedenberg, K. and Mehler, E. L., *Adv. Quantum Chem.*, **5**, 27 (1970).

Gallup, G. A., *J. Chem. Educ.*, **65**, 671 (1988).

McWeeney, R., *Coulson's Valence*, Oxford (1979).

Murrell, J. N., Kettle S. F. A. and Tedder, J. M., *The Chemical Bond*, 2nd edition, John Wiley & Sons (1978).

Murrell, J. N., Kettle S. F. A. and Tedder, J. M., *Valence Theory*, 2nd edition, John Wiley & Sons (1974).

Slater, J. C., *Quantum Theory of Atomic Structure*, Volumes 1, 2, McGraw-Hill (1960).

Slater, J. C., *Quantum Theory of Molecules and Solids*, Volumes 1, 2, McGraw-Hill (1963).

2 What is the Basis of the Model we use to Describe the Orbital Structure of Molecules?

Many of our discussions use the idea of a secular determinant to derive energy levels in molecules and solids. Where does it come from?

Actually, our discussion here is a rather standard treatment[1,2] of this topic. We should include it since many of the results we use elsewhere have their origin here as you noted. First though, we should point out that virtually all of the material in this book centers around the simplest electronic models of all, namely one-electron ones. Many questions, which we don't tackle in our discussions (the difficult ones), involve numerical aspects of chemistry, such as bond energies, reaction pathways, aspects of metal–insulator transitions etc., and these are only accessible using numerically more exact methods.

We begin with the idea of representing the molecular wavefunction (the molecular orbitals or MOs ψ_j) in terms of a linear combination of atomic orbitals or AOs, ϕ_i, the LCAO approximation.

$$\psi_j = \sum_i c_{ij} \phi_i \tag{2.1}$$

This is quite a simple and broad approach. The problem however, is to evaluate the mixing coefficients, the c_{ij}. To do this we need to evaluate the energy of the system described by such a collection of MO's and to find the set of coefficients that gives the lowest energy for the orbital. This relies on the Variational Theorem which says that any approximate wavefunction will always give an energy that is higher than the true energy. Recall that the energy is simply given by

$$E = \frac{\langle \psi | \mathscr{H}^{\text{eff}} | \psi \rangle}{\langle \psi^2 \rangle} \tag{2.2}$$

Here we have used some effective Hamiltonian, $\mathscr{H}^{\text{eff}}$, to describe the quantum mechanical problem at hand. (We will talk a little more about this later.) To obtain the lowest energy we need to minimize equation (2.2) with respect to the c_{ij}. Namely

$$\frac{\partial E}{\partial c_i} = 0 \qquad \text{for all } i \tag{2.3}$$

where we have ignored the MO labels for the moment. Expanding equation (2.2)

$$E = \frac{\left\langle \left(\sum_i c_i \phi_i \right) | \mathscr{H}^{\text{eff}} | \left(\sum_i c_i \phi_i \right) \right\rangle}{\left\langle \left(\sum_i c_i \phi_i \right)^2 \right\rangle}$$

$$= \frac{\sum_i \sum_j c_i c_j \langle \phi_i | \mathscr{H}^{\text{eff}} | \phi_j \rangle}{\sum_i \sum_j c_i c_j \langle \phi_i | \phi_j \rangle} \tag{2.4}$$

Let's introduce the terms and $H_{ij} = \langle \phi_i | \mathcal{H}^{\text{eff}} | \phi_j \rangle$ and $S_{ij} = \langle \phi_i | \phi_j \rangle$. When $i \neq j$, H_{ij} is the interaction integral (sometimes called the resonance or hopping integral), between the atomic orbitals $\phi_{i,j}$, and S_{ij} is the overlap integral between them. For $i = j$, H_{ii} is called the Coulomb or on-site integral. Thus equation (2.4) becomes

$$E = \frac{\sum_i \sum_j c_i c_j H_{ij}}{\sum_i \sum_j c_i c_j S_{ij}} \tag{2.5}$$

or

$$E \sum_i \sum_j c_i c_j S_{ij} = \sum_i \sum_j c_i c_j H_{ij} \tag{2.6}$$

Differentiating with respect to a specific coefficient, say c_m, leads to

$$E \sum_j c_j S_{mj} + E \sum_i c_i S_{mi} = \sum_j c_j H_{mj} + \sum_i c_i H_{mi} \tag{2.7}$$

Now we can appreciate that $S_{ij} = S_{ji}$ (although we will not prove it), and so $H_{ij} = H_{ji}$ such that

$$E \sum_i c_i S_{mi} = \sum_i c_i H_{mi} \tag{2.8}$$

or with a little rearrangement

$$\sum_i c_i (H_{mi} - E S_{mi}) = 0 \tag{2.9}$$

which may be rewritten in a fuller form as

$$\begin{matrix}
(H_{11} - S_{11}E)c_1 & (H_{12} - S_{12}E)c_2 & \ldots & (H_{1n} - S_{1n}E)c_n = 0 \\
(H_{21} - S_{21}E)c_1 & (H_{22} - S_{22}E)c_2 & \ldots & (H_{2n} - S_{2n}E)c_n = 0 \\
\vdots & \vdots & \vdots & \vdots \\
(H_{n1} - S_{n1}E)c_1 & (H_{n2} - S_{n2}E)c_2 & \ldots & (H_{nn} - S_{nn}E)c_n = 0
\end{matrix} \tag{2.10}$$

or in matrix form as

$$\begin{pmatrix}
H_{11} - S_{11}E & H_{12} - S_{12}E & \ldots & H_{1n} - S_{1n}E \\
H_{21} - S_{21}E & H_{22} - S_{22}E & \ldots & H_{2n} - S_{2n}E \\
\vdots & \vdots & \vdots & \vdots \\
H_{n1} - S_{n1}E & H_{n2} - S_{n2}E & \ldots & H_{nn} - S_{nn}E
\end{pmatrix}
\begin{pmatrix}
c_1 \\
c_2 \\
\vdots \\
c_n
\end{pmatrix} = 0 \tag{2.11}$$

The first matrix represents the pairwise interactions of all of the orbitals of the problem. In order for the set of equations of (2.10) to be self-consistent, then the following determinant needs to be zero:

$$\begin{vmatrix}
H_{11} - S_{11}E & H_{12} - S_{12}E & \ldots & H_{1n} - S_{1n}E \\
H_{21} - S_{21}E & H_{22} - S_{22}E & \ldots & H_{2n} - S_{2n}E \\
\vdots & \vdots & \vdots & \vdots \\
H_{n1} - S_{n1}E & H_{n2} - S_{n2}E & \ldots & H_{nn} - S_{nn}E
\end{vmatrix} = 0 \tag{2.12}$$

or

$$|\mathbf{H} - \mathbf{S}E| = 0 \tag{2.13}$$

where $\mathbf{H}$ (the Hamiltonian matrix) is a matrix containing all of the H_{ij} and $\mathbf{S}$ a matrix containing all of the S_{ij}. Equation (2.12) is called the secular determinant. Its solution leads to values of the E_i (the eigenvalues of $\mathbf{H}$) and subsequent substitution into equation (2.10) leads to values of the c_{ij} (the eigenvectors of $\mathbf{H}$).

How do you estimate numerical values of the terms in equation (2.12), or does one need to?

Let's begin by exploring ways of setting numerical values for the off- diagonal elements H_{ij}. The most common method is to use an approach due to Robert Mulliken but called the Wolfsberg–Helmholz equation after two[3] of its early proponents. This is to put H_{ij} equal to either the scaled arithmetic or geometric mean of the H_{ii} values of the atomic orbitals. This is shown in equation (2.14):

$$H_{ij} = (1/2)KS_{ij}(H_{ii} + H_{jj})$$
$$H_{ij} = KS_{ij}(H_{ii}H_{jj})^{1/2} \tag{2.14}$$

K is a scaling constant usually put equal to 1.75. Thus the strength of the interaction between two orbitals depends (logically) on the overlap between them. Now $H_{ii} = \langle \phi_i | \mathscr{H}^{\text{eff}} | \phi_i \rangle$, is just the energy of an electron in the atomic orbital ϕ_i. The value of this for use in one-electron models can be chosen in several ways, but perhaps the best is as the configurationally averaged atomic ionization potential (see our discussion at the end of this chapter) which leads to the energies $e_{\text{s,p,d}}$. These are negative which means that H_{ii} is always negative and H_{ij} negative if S_{ij} is positive. The S_{ij} can be evaluated numerically for a given problem by choosing Slater-type orbitals on each center. The particular parameterization we have described is that appropriate for the extended Hückel method[4] and has been extremely useful in the study of the orbital problems of a multitude of organic, inorganic, organometallic and solid-state problems.

It is interesting that we have never actually defined the Hamiltonian, $\mathscr{H}^{\text{eff}}$, for this problem. All that are important are the values, H_{ii} and H_{ij} of its matrix elements.

Yes, this is indeed true. There is, however, an even simpler model derived from equation (2.12) that has been of very general use for sixty years. This was devised by Erich Hückel[5] for the π systems of organic molecules. It enables the orbital energies of these molecules to be expressed in terms of only two parameters. (Erich's brother Walter worked hard for many years to persuade the German chemical community to accept his brother's ideas. For a long time the approach was more widely used outside Germany than inside.)

Although the Hückel model was for many years used exclusively for the study of the π systems of organic molecules, the approach has a much broader usage. It does make some grand simplifications. First, it sets all the overlap integrals (S_{ij}) equal to zero, unless $i = j$ in which case $S_{ii} = 1$. Second, it introduces a notation, $H_{ii} = \alpha_i$ and $H_{ij} = \beta_{ij}$. Thus equation (2.11) becomes

$$\begin{pmatrix} \alpha_1 - E & \beta_{12} & \cdots & \beta_{1n} \\ \beta_{12} & \alpha_2 - E & \cdots & \beta_{2n} \\ \vdots & \vdots & \vdots & \vdots \\ \beta_{1n} & \beta_{2n} & \cdots & \alpha_n - E \end{pmatrix} \begin{pmatrix} c_1 \\ c_2 \\ \vdots \\ c_n \end{pmatrix} = 0 \tag{2.15}$$

and equation (2.12) becomes

$$\begin{vmatrix} \alpha_1 - E & \beta_{12} & \cdots & \beta_{1n} \\ \beta_{12} & \alpha_2 - E & \cdots & \beta_{2n} \\ \vdots & \vdots & \vdots & \vdots \\ \beta_{1n} & \beta_{2n} & \cdots & \alpha_n - E \end{vmatrix} = 0 \qquad (2.16)$$

Second, it sets all $H_{ij} = 0$ unless the atoms are bonded through the σ framework. So, depending on the way the atoms are connected, the β_{ij} will either be set equal to zero or to β, dropping the subscripts. If all the orbitals are of the same type (e.g., $p\pi$) and lie on the same type of atom (e.g., C) then all the α_i will be equal. So for the π system of cyclic $C_3H_3{}^+$, Structure **2.1**, the secular determinant is

$$\begin{vmatrix} \alpha - E & \beta & \beta \\ \beta & \alpha - E & \beta \\ \beta & \beta & \alpha - E \end{vmatrix} = 0 \qquad (2.17)$$

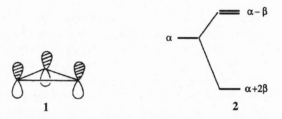

with roots $E = \alpha + 2\beta$ and $E = \alpha - \beta$ (twice). Thus the molecular orbital diagram is given by Structure **2.2**.

How do you get the character of the molecular orbitals, namely the c_i, from this?

We can evaluate the form of the orbitals, the c_i, by substitution of these energies into equation (2.15). So, for example, for the MO for which $E = \alpha + 2\beta$

$$\begin{pmatrix} 2\beta & \beta & \beta \\ \beta & 2\beta & \beta \\ \beta & \beta & 2\beta \end{pmatrix} \begin{pmatrix} c_1 \\ c_2 \\ c_3 \end{pmatrix} = 0 \qquad (2.18)$$

which after some arithmetic gives $c_1 = c_2 = c_3 = 1$. This is a simple example but you can readily do this by hand for smaller systems. A computer program will give you the answer more quickly.

For the open chain molecule allyl (Structure **2.3**) there is now no entry in the Hamiltonian matrix in the 1,3 position and so the secular determinant is as in equation (2.19). Its solution gives the energy levels of Structure **2.4**:

$$\begin{vmatrix} \alpha - E & \beta & 0 \\ \beta & \alpha - E & \beta \\ 0 & \beta & \alpha - E \end{vmatrix} = 0 \qquad (2.19)$$

The general solution[1,2] for a polyene chain of N carbon atoms $((CH)_N H_2)$ is

$$E(j) = \alpha + 2\beta \cos(j\pi/N + 1) \qquad (2.20)$$

where j is a quantum number labeling the jth level. As Figure 2.1 shows for the first few polyenes, the lowest energy level is always bonding between all nearest neighbors. As the stack of levels is climbed, the energy increases and the number of nodes increases just as in

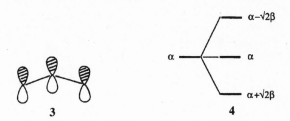

3

4

the case of atomic orbitals. Of course exactly the same arguments would apply to the (hypothetical) case of a chain of hydrogen atoms where the interactions are between the 1s orbitals of hydrogen rather than between the $2p\pi$ orbitals of carbon.

There must be many generalities concerning the form of these energy level diagrams.

Yes. We have already shown the general expression for the one-dimensional chain in equation (2.20). Another applies to systems where each atom is linked to every other atom. Triangular H_3 or C_3H_3 and tetrahedral H_4 are the two simplest examples. In these N-atomic species where each atom is Γ-coordinate there is a single bonding orbital at $E = \alpha + \Gamma\beta$ and $N - 1$ antibonding orbitals at $E = \alpha + (\Gamma\beta/(N - 1))$. Perhaps the best known examples however, are those discussed in Chapter 13 for cyclic systems. Here (Structure **2.5**) the lowest energy orbital always lies at an energy of $\alpha + 2\beta$ and all the others then lie in pairs (except for the highest lying orbital of even-membered rings that lies at an energy of $\alpha - 2\beta$). It turns out that using the cyclic group of order n (C_n) to describe the symmetry properties of these molecules then each irreducible representation is used once, so each level has a different symmetry label.

It is interesting that you have used the H_3 system in the same sentence as C_3H_3 in the same manner as earlier when comparing energies of linear hydrogen and $(CH)_N$ chains. This is really using the concept of a 'topological' theory.

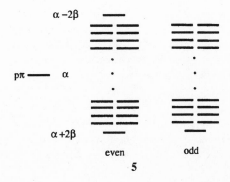

5

E

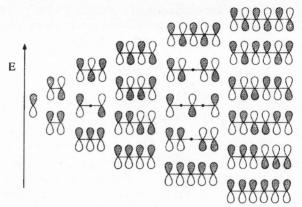

Figure 2.1
The trends in the energy levels of the first few polyenes

Yes, the energy levels of cyclic H_n systems (only one is known experimentally) are exactly the same as those for the $p\pi$ orbitals of the corresponding cyclic polyenes. The Hückel α and β now apply to σ hydrogen 1s–1s interactions rather than carbon $2p\pi$– $2p\pi$.

How do you apply these Hückel ideas to heteronuclear systems?

These rather simple ideas do give insights as to how orbitals interact in more complex molecules. Let's take the system where the two orbitals are of different H_{ii} or different electronegativity. The Hückel secular determinant is then

$$\begin{vmatrix} \alpha_1 - E & \beta \\ \beta & \alpha_2 - E \end{vmatrix} = 0 \tag{2.21}$$

By expansion of the secular determinant we can show that it has roots with energies $E \sim [(\alpha_1 + \alpha_2) \pm (\alpha_1 - \alpha_2)^2 + 4\beta^2)^{1/2}]/2$. Expanding under the square root and taking the leading term gives

$$E \sim \alpha_1 + \beta^2/(\alpha_1 - \alpha_2) \quad \text{and} \quad \alpha_2 - \beta^2/(\alpha_1 - \alpha_2) \tag{2.22}$$

These are bonding and antibonding orbitals associated with ϕ_1 and ϕ_2. The stabilization energy of the bonding orbital is just $\varepsilon = \beta^2/(\alpha_1 - \alpha_2)$. (We will see an expression of the same type again in Chapter 7.)

This simple result shows that the interaction or stabilization energy between two orbitals depends on the square of the interaction integral, β, between them and depends inversely on their energy separation. Thus the closer two atomic orbitals are in energy, the larger is their interaction. In a more obvious way, since β depends on the overlap integral, the larger S, the larger is the interaction too.

Exactly. This tells us immediately why, when constructing the molecular orbitals of methane, we do not need to consider the carbon 1s orbitals. On both overlap grounds (these orbitals are quite contracted and so H_{ij} is close to zero) and from energy gap considerations (carbon 1s and hydrogen 1s are far apart in energy) the interactions are negligible.

Let's see now what the wavefunction looks like for the deeper-lying molecular orbital from equation (2.21). Anticipating our result we will write it as $\psi = \phi_1 + a\phi_2$. Then the energy of the orbital is just

$$E \approx \langle \phi_1 | \mathscr{H}^{\text{eff}} | \phi_1 \rangle + 2a\langle \phi_1 | \mathscr{H}^{\text{eff}} | \phi_2 \rangle + a^2 \langle \phi_2 | \mathscr{H}^{\text{eff}} | \phi_2 \rangle$$
$$= \alpha_1 + 2a\beta + a^2\alpha_2 \tag{2.23}$$

Since $a < 1$ we ignore the small term containing a^2. Then comparison with the expression in equation (2.21), $2a\beta = \beta^2/(\alpha_1 - \alpha_2)$ and $a \approx \beta/(\alpha_1 - \alpha_2)$. Thus the stronger the interaction between two orbitals, either in terms of the interaction integral between them, β, or their energy separation $(\alpha_1 - \alpha_2)$, the larger the coefficient of the upper atomic orbital in the lower molecular orbital. This means that the further the two interacting orbitals are apart in energy, the more the lower energy molecular orbital will look like the lower energy starting atomic orbital. Similar considerations apply to the higher energy orbital of course.

This discussion reminds me of the idea of electronegativity. How does this connect with the orbital picture?

There has been a large amount of effort expended over the years to clarify and quantify the concept of electronegativity (χ). This is almost certainly because of its rather vague initial definition combined with a sense that it is a very useful notion. Linus Pauling[6] defined it as 'the power of an atom in a molecule to attract electrons to itself.' His numerical scheme, if you remember, related the electronegativity difference to the excess bond energy in AB molecules, Δ, as

$$\chi_A - \chi_B = K\Delta^{1/2}$$

where D_{XY} is the XY bond energy and

$$\Delta = D_{AB} - \sqrt{D_{AA}D_{BB}} \tag{2.24}$$

He set $K = 23^{-1/2}$ that allowed a scale lying between ~ 0 and ~ 4 and set its origin by putting $\chi_F = 4.0$.

Robert Mulliken defined electronegativity as the mean of the ionization potential and electron affinity. The origin of his approach is simple to see. Examination of the molecular orbital diagram for HF of Figure 6.1 shows that the bonding orbital is composed of both a hydrogen and fluorine character. We have just shown that the relative magnitude of a hydrogen atom character in the bonding orbital depends on the inverse of the energy separation (ΔE) of the starting orbitals, here fluorine, 2p and hydrogen, 1s atomic orbitals, i.e., roughly the difference in their ionization potential. If we write this bonding orbital as

$$\psi = a\phi_H + b\phi_F \tag{2.25}$$

then the ratio $b/a > 1$, since the bonding orbital lies closer energetically to the starting fluorine atomic orbital. This ratio becomes larger as the energy separation, or difference in Mulliken electronegativity, increases. Thus the electron density will be more localized on fluorine, the larger b/a. (If the energy separation is very large then the orbital will be virtually a pure fluorine 2p orbital.)

It's interesting that this is a result exactly in the spirit of Pauling's original definition but with an orbital formulation.

Yes. Another related route is from the values of the ionization energies determined experimentally. This is shown in equation (2.26)

$$\chi = \frac{n_s e_s + n_p e_p}{n_s + n_p}$$

(2.26)

and gives the electronegativity, as a weighted sum of the configurationally averaged atomic ionization potentials, $e_{s,p}$. What do we mean by configurationally averaged? In the many-electron atom there will be several states associated with a given electron occupation of (say) the valence p orbitals by more than one electron. They differ in energy because of different Coulomb and exchange terms. For example, one may be a singlet state, another a triplet. By weighting the energies from all these states, an average energy is generated. An electronegativity scale of this type has been constructed by Allen[7] and is shown pictorially in Figure 2.2. It is perhaps, just a more modern version of the Mulliken formulation. (Shown in Table 2.1 are numerical values of Allen's scale alongside those of Mulliken and Pauling.)

There are several other ways to approach the idea of electronegativity. One comes from the idea of the pseudopotential of Figure 4.2. Since the ionization energy is expected to scale with $(1/r)_{av}$ where r is the electron–nucleus distance, then it should scale with the reciprocal of the pseudopotential radius $1/R_c$ (and does[8]). Thus we may write

$$\chi = c_1(1/r_s) + c_2(1/r_p)$$

(2.27)

where r_s and r_p are the s and p orbital radii (the R_c of Figure 4.2 for the different orbitals). The c_i are weighting coefficients.

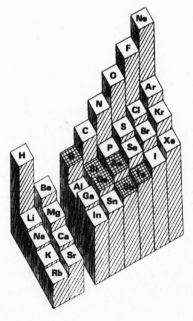

Figure 2.2
The Allen electronegativities of the elements. (Reproduced by permission of The American Chemical Society from reference 7)

Table 2.1 Electronegativities of the elements comparing the scales of Pauling, Mulliken and Allen

Atom	χ_P	χ_M	χ_A	Atom	χ_P	χ_M	χ_A
H	2.20	3.06	2.30	Ga	1.81	1.34	1.76
Li	0.98	1.28	0.91	Ge	2.01	1.95	1.99
Be	1.57	1.99	1.58	As	2.18	2.26	2.21
B	2.04	1.83	2.05	Se	2.55	2.51	2.42
C	2.55	2.67	2.54	Br	2.96	3.24	2.69
N	3.04	3.08	3.07	Kr		2.98	2.97
O	3.44	3.21	3.61	Rb	0.82	0.99	0.71
F	3.98	4.42	4.19	Sr	0.95	1.21	0.96
Ne		4.60	4.79	In	1.78	1.30	1.66
Na	0.93	1.21	0.87	Sn	1.96	1.83	1.82
Mg	1.31	1.63	1.29	Sb	2.05	2.06	1.98
Al	1.61	1.37	1.61	Te	2.10	2.34	2.16
Si	1.90	2.03	1.92	I	2.66	2.88	2.36
P	2.19	2.39	2.25	Xe	2.60	2.59	2.58
S	2.58	2.65	2.59	Cs	0.79		
Cl	3.16	3.54	2.87	Ba	0.89		
Ar		3.36	3.24	Tl	2.04		
K	0.82	1.03	0.73	Pb	2.33		
Ca	1.00	1.30	1.03	Bi	2.02		

There are also interesting connections between the way electronegativity is described here and the idea of hard and soft acids and bases[9,10] that are well worth reading.

References

1. Streitweiser, A., *Molecular Orbital Theory for Organic Chemists*, John Wiley & Sons (1961).
2. Heilbronner, E. and Bock, H., *The HMO Model and its Applications*, John Wiley & Sons (1976).
3. Wolfsberg, M. and Helmholz, L., *J. Chem. Phys.*, **20**, 837 (1952); McGlynn, S. P., Vanquickenborne, L. G., Kinoshita, M. and Carroll, G. G. (eds), *Introduction to Applied Quantum Chemistry*, Holt, Rheinhart and Winston (1972).
4. Hoffmann, R., *J. Chem. Phys.*, **39**, 1397 (1962).
5. Hückel, E., *Z. Phys.*, **70**, 204 (1931).
6. Pauling, L., *The Nature of the Chemical Bond*, Third Edition, Cornell University Press (1960).
7. Allen, L. C., *J. Am. Chem. Soc.*, **111**, 9003 (1989).
8. Zunger, A., in *Structure and Bonding in Crystals*, Vol. 1, O'Keeffe, M. and Navrotsky, A. (eds), Academic Press (1981).
9. Parr, R. G. and Pearson, R. G., *J. Am. Chem. Soc.*, **105**, 7512 (1983).
10. Pearson, R. G., *J. Chem. Educ.*, **64**, 561 (1987); *Inorg. Chem.*, **23**, 734 (1988).

General References and Further Reading

Dewar, M. J. S. and Dougherty, R. C., *The PMO Theory of Organic Chemistry*, Plenum (1975).

Dewar. M. J. S., *Theory of Molecular Orbitals*, McGraw-Hill (1969).

Hehre, W. J., Radom, L., Schleyer, P. v R and Pople, J. A., *Ab Initio Molecular Orbital Theory*, John Wiley & Sons (1986).

Hout, R. F., Pietro, W. J. and Hehre, W. J., *A Pictorial Approach to Molecular Structure and Reactivity*, John Wiley & Sons (1984).

Kutzelnigg, W., *Angew. Chem.*, **35**, 573 (1996).

McWeeney, R., *Methods of Molecular Quantum Mechanics*, Academic Press (1989).

Pearson, R. G., *Accts. Chem. Res.*, **23**, 1 (1993).

Pople, J. A. and Beveridge, δ. δ., *Approximate Molecular Orbital Theory*, McGraw-Hill (1970).

Richards, W. G. and Cooper, D. L., *Ab Initio Molecular Orbital Calculations for Chemists*, 2nd Edition, Oxford University Press (1982).

Sanderson, R. T., *Chemical Bond and Bond Energy*, Academic Press (1976).

Webster, B., *Chemical Bonding Theory*, Blackwell (1990).

Yates, K., *Hückel Molecular Orbital Theory*, Academic Press (1978).

3 sp³ Hybrids and the Molecular Orbitals of Methane. What's the Difference?

There seems to be quite a difference in the type of theoretical treatment of the chemical bonding in main group diatomic molecules (perhaps our first introduction to molecular orbital theory) and that encountered in the study of organic chemistry. Although we come across molecular orbitals when considering the Woodward–Hoffmann rules[1], our first exposure to chemical bonding in organic molecules is in terms of the sp^3 hybrids used[2] to view that quintessential organic molecule, methane. We usually view this species as being built from the overlap of four hydrogen 1s orbitals with the orbitals of a carbon atom that is 'prepared for bonding', i.e., with four sp^3 hybrid orbitals directed toward the corners of a tetrahedron. The molecular orbital picture is quite different. Figures 3.1 and 3.2 show these two ways of looking at the bonding in the molecule. The molecular orbitals are clearly 'delocalized' over the molecule but the hybrid orbitals 'localized' between a bonded pair of atoms. One observation that is immediately perplexing is that in the T_d point group we know that the highest

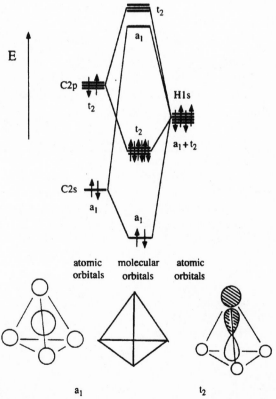

Figure 3.1
The molecular orbital picture for methane (only one component of the t_2 set is shown)

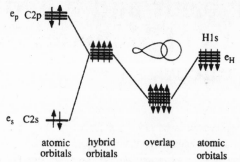

Figure 3.2
The localized picture for methane in terms of four sp³ hybrid orbitals (only one is shown) directed toward the corners of a tetrahedron

degeneracy possible is three ($t_{1,2}$) so that there surely cannot be four equivalent hybrid energy levels with the same energy each describing a bond. Obviously the 'organic chemists' approach,' can't be all that bad though, since pushing electrons around in molecules to view chemical reactions electronically has been used for many years and is a very useful model. It is too a natural way of using Lewis' ideas in both organic and inorganic chemistry. How are the two related?

One of the obvious conceptual drawbacks of the molecular orbital approach (equation (1.5)) is that the molecular orbitals themselves are delocalized over (in principle) many or all of the atoms of the molecule. In large molecules, this can be a large number of orbitals. Therefore one of the advantages of the localized orbital approach, as you noted, is that one can often associate a pair of electrons with a particular chemical bond as in a Lewis structure. There are several observations that encourage us to do so. For example, the length of a C–H bond in hydrocarbons is always around 1.08 Å, its stretching force constant around 500 N m⁻¹, and its bond energy about 405 kJ/mol. Similar constancy of bond parameters is found for C–C bonds in unstrained environments. For example, addition of a CH_2 unit to a saturated hydrocarbon chain leads to an increase in the enthalpy of formation of 346 kJ/mol (for the carbon) + 2 × 411 kJ/mol (for the hydrogens) = 1168 kJ/mol. This algorithm leads to an error of only one or two percent when compared with experiment. These are not results which are obvious from the molecular orbital model. Your question as to which is correct is a good one. Both are in fact 'correct' but we will see that some stringent limitations apply to the use of the localized model and the way you have drawn Figure 3.2.

Let's start with, not methane, but the molecule BeH_2, assumed to be linear. Here the conventional localized description of this molecule would be as in Structure **3.1**. Two pairs of electrons occupy localized Be–H bonding orbitals constructed from the central atom 2s and 2p orbitals, namely two sp hybrid orbitals. Figure 3.3(a) shows the molecular orbital diagram for BeH_2. Just like that of Figure 3.1, it's a simple example of the technique[3] of

H:Be:H

1

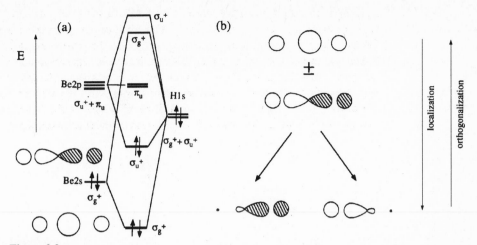

Figure 3.3
(a) A molecular orbital diagram for linear BeH$_2$. The forms of the σ_g^+ and σ_u^+ orbitals are shown; (b) a linear combination of the two occupied orbitals leads to two localized orbitals. Orthogonalization of these two orbitals leads back to the pair of occupied molecular orbitals

assembly of a molecular orbital diagram. First one identifies the irreducible representations of the central atom orbitals at left, then those of the ligand orbitals at right, and then constructs an energy level diagram by identifying pairs of orbitals of the same symmetry species. The energies of the starting orbitals are those of the atomic s and p orbitals, $e_{s,p}$.

An important point is that in both occupied molecular orbitals, the wavefunction is distributed over all three atoms, typical molecular orbitals as we described earlier. There are two occupied bonding molecular orbitals and two Be—H bonds, so the average Be—H bond order (using this term in its simplest sense) is one.

However, there is more to it than that. It is interesting to see that a judiciously chosen linear combination of these two occupied orbitals leads to two functions, shown in Figure 3.3(b), which are mirror images of each other, and look just like the 'localized' orbitals that one would use to describe this molecule. They are frequently called bond orbitals[4].

But they are not orthogonal.

Quite true. Neither do they transform as any of the irreducible representations of the molecular point group (D$_{\infty h}$). They can be orthogonalized but the result is just the reverse of the localization process and production of the same pair of molecular orbitals we started with (Figure 3.3(b)).

A similar picture is found for the methane molecule. Figures 3.1 and 3.2 show molecular orbital and localized pictures. The molecular orbital diagram shows a trio of bonding orbitals (symmetry species t$_2$) which involve no carbon 2s character at all, and a single orbital (a$_1$), lying deeper in energy, which involves no carbon 2p.

It is interesting that this picture does not violate the rule that the wavefunctions must obey the symmetry properties of the point group.

Yes. Structure **3.2** shows how the localized orbitals (we show just one, derived from one t_2 component) can be constructed from the delocalized ones in just the same fashion as for BeH_2. Since, as you noted, group theory tells us that there cannot be a set of four degenerate orbitals in a molecule with this geometry, this localized picture as drawn cannot be correct from this point of view. Added weight comes from photoelectron spectroscopy. Figure 3.4 shows the photoelectron spectrum of methane[5]. What is clear to see is one peak close to the energy of the carbon 2s orbital (the a_1 orbital of the molecular orbital picture) and one close to the energy of a hydrogen 1s orbital, with greater intensity (the t_2 orbitals).

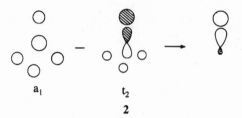

$$a_1 \qquad\qquad t_2$$
$$2$$

So experimental data shows that the molecular orbital picture reigns triumphant and the localized view of Figure 3.2 is exposed as incorrect.

Well, not quite, but it does tell us that we need to proceed cautiously with the localized picture when viewing photoelectron spectra. The solution to this conundrum is found when we write down[4] the wavefunction for the ground electronic state of the molecule as a whole, as was first shown by John Lennard-Jones. Let's start with the simplest system with two electrons (1,2) and two orbitals. This might be the case for methylene shown in Structure **3.3**. We will need to include the electron spin ($\pm\frac{1}{2}$ or the labels α, β) in the wavefunction as well as the spatial part, $\xi_{1,2}$ of the two molecular orbitals that house the two electrons. The localized picture is shown in Structure **3.4**. As before in BeH_2, localized orbitals may be constructed by taking the linear combinations $\xi_1 \pm \xi_2$. We could write a wavefunction for this state as

$$\Psi = \xi_1(1)\alpha(1)\xi_2(2)\beta(2) \tag{3.1}$$

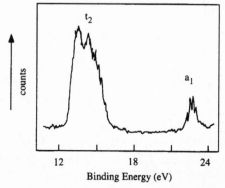

Figure 3.4
The photoelectron spectrum of methane. (The relative intensities of the two peaks depends upon the wavelength of the ionizing radiation, via the photoelectric cross-sections)

3   **4**

but this is just as good as

$$\Psi = \xi_1(2)\alpha(2)\xi_2(1)\beta(1) \tag{3.2}$$

However, according to the Pauli Principle the total wavefunction must be antisymmetric with respect to electron exchange (i.e., must change sign). So we must write for the total wavefunction Ψ

$$\Psi = \xi_1(1)\alpha(1)\xi_2(2)\beta(2) - \xi_1(2)\alpha(2)\xi_2(1)\beta(1) \tag{3.3}$$

or

$$\Psi = \begin{vmatrix} \xi_1(1)\alpha(1) & \xi_2(1)\beta(1) \\ \xi_1(2)\alpha(2) & \xi_2(2)\beta(2) \end{vmatrix} \tag{3.4}$$

You can easily test to see that by exchanging the electrons (swapping the labels 1,2 via the operator $\mathscr{P}_{12}$) the total wavefunction changes sign.

$$\begin{aligned}
\mathscr{P}_{12}\Psi &= \xi_1(2)\alpha(2)\xi_2(1)\beta(1) - \xi_1(1)\alpha(1)\xi_2(2)\beta(2) \\
&= -\xi_1(1)\alpha(1)\xi_2(2)\beta(2) + \xi_1(2)\alpha(2)\xi_2(1)\beta(1) \\
&= -[\xi_1(1)\alpha(1)\xi_2(2)\beta(2) - \xi_1(2)\alpha(2)\xi_2(1)\beta(1)].
\end{aligned} \tag{3.5}$$

Equation (3.4) is called a Slater determinant and represents an antisymmetrized product wavefunction. Writing products such as $\xi_j(i)\beta(i)$ as spin orbitals in a simpler notational way, that is to say as $\psi_{a,b}(i)$ then

$$\Psi = \begin{vmatrix} \psi_a(1) & \psi_b(1) \\ \psi_a(2) & \psi_b(2) \end{vmatrix} \tag{3.6}$$

Sometimes we shorten this to $\mathscr{A}(\psi_a\bar{\psi}_a\psi_b\bar{\psi}_b)$ where $\mathscr{A}$ is an antisymmetrizer or just $|\psi_a\bar{\psi}_a\psi_b\bar{\psi}_b|$ where the bar in both cases represents an electron of opposite spin.

So we can see from this expression one way of viewing the Pauli Principle. If the two electrons are described by the same spin orbital, i.e., $\psi_a = \psi_b$, then two columns of the Slater determinant are identical, a situation that leads to the total wavefunction being zero, a situation that is not allowed. Thus the electrons in an atom or molecule must be described by different spin orbitals, or alternatively two electrons with the same spin may not share the same atomic or molecular orbital.

Yes, this is quite a clear way to show this. Now let's see what happens as we change the basis functions, the $\xi_j(i)$ or $\psi(i)$ that go into the secular determinant. First we need to note a fundamental property of determinants shown in equation (3.7).

$$\begin{vmatrix} a & b \\ c & d \end{vmatrix} = \begin{vmatrix} a & (\lambda a + b) \\ c & (\lambda c + d) \end{vmatrix} = ad - bc \tag{3.7}$$

This holds for all λ. Making use of this trick allows us a series of simple manipulations:

$$
\begin{aligned}
\Psi &= \begin{vmatrix} \psi_a(1) & \psi_b(1) \\ \psi_a(2) & \psi_b(2) \end{vmatrix} \\[2mm]
&= \begin{vmatrix} \psi_a(1)+\psi_b(1) & \psi_b(1) \\ \psi_a(2)+\psi_b(2) & \psi_b(2) \end{vmatrix} \\[2mm]
&= \begin{vmatrix} [\psi_a(1)+\psi_b(1)] & \tfrac{1}{2}[\psi_b(1)-\psi_a(1)] \\ [\psi_a(2)+\psi_b(2)] & \tfrac{1}{2}[\psi_b(2)-\psi_a(2)] \end{vmatrix} \\[2mm]
&= -\tfrac{1}{2}\begin{vmatrix} [\psi_a(1)+\psi_b(1)] & [\psi_a(1)-\psi_b(1)] \\ [\psi_a(2)+\psi_b(2)] & [\psi_a(2)-\psi_b(2)] \end{vmatrix}
\end{aligned}
\tag{3.8}
$$

Apart from a new normalization term, we have converted a state described by the basis orbitals ψ_a and ψ_b into one described by $\psi_a+\psi_b$ and $\psi_a-\psi_b$. Thus for methylene we have immediately generated a localized picture with one electron each in $\xi_1+\xi_2$ and $\xi_1-\xi_2$ from the delocalized picture, with one electron in ξ_1 and one in ξ_2. As we can see, as far as the *total* wavefunction for the ground electronic state is concerned, the two are completely equivalent. This two-electron problem was done for simplicity. You can work a little harder to do this for BeH_2 or CH_4.

Is there anything special about the functions $\psi_a \pm \psi_b$ or $\xi_1 \pm \xi_2$?

No, the functions $\psi_a \pm \psi_b$ or $\xi_1 \pm \xi_2$ have no special properties *per se*. They are just an arbitrary choice as we can see from the manipulation of equation (3.8). The ones that we have chosen though are 'equivalent' orbitals and have the chemical advantage that they point towards the hydrogen atoms where we anticipate there to be a chemical bond. The formal definition of an equivalent set of orbitals is that one may be transformed into another by a symmetry operation of the molecular point group. Although the choice is arbitrary, we usually select the bond or hybrid orbitals so that the mutual electrostatic repulsion of the electrons between them is minimized. (See the discussion of the VSEPR scheme of Chapter 17.) There are some other ways of deciding where the 'bonds' are in molecules[6,7] which are very much worth studying.

So, in terms of the total wavefunction, the molecule is oblivious to the use of a delocalized basis set or a localized one derived from it. But when can we use either approach and when do we have to use one in particular?

For any observable for which all of the electrons need to be included, the use of either approach will be a valid one from this discussion. Two examples might be the total energy and the electron density. It is the latter, of course, which is the primary feature of importance in looking at chemical bonds in terms of 'electron pushing.' Either approach will accordingly be a valid descriptor of the system for all-electron, or 'collective,' properties[8]. Photoelectron spectra however, do not fall into this category (a single electron is excited) and this is where the localized model is inapplicable. Here we must use the delocalized model with its molecular orbital structure to understand such results. This leads to the results of Figure 3.4. Thus, in a way, the diagram of Figure 3.2 is just inappropriately drawn when it comes to looking at such physical observations. The use of sp³ hybrid orbitals, and other hybrid orbital constructs in general though, are perfectly good ways to view chemical bonding in molecules and solids.

Are there any other ways to probe this question experimentally?

There are some very nice photoelectron experiments that further strengthen the molecular orbital picture of Figure 3.1 that have been carried out by Christopher Brion[9]. Imagine a pool table with a cover over a part of its length. Underneath the cover is a ball executing some sort of motion. How can we determine its behavior without lifting the cover and looking? One way is to shoot in a second ball with known momentum and then measure the velocities of the two balls as they emerge from under the cover after having collided. (This idea requires a friction-free pool table.) In atoms and molecules one can do the same thing by firing an electron at the molecule and then measuring the momenta of the pair of electrons as they leave the atom. This is known as e–$2e$ spectroscopy (one electron in, two out) after the nomenclature of high-energy physics. By fixing the energetics to study the deepest bound electrons of Figure 3.1, one finds that the electron is indeed moving in a spherical (i.e., 's') orbit around the carbon nucleus. Similarly a 'p'-like motion is found for the less tightly bound electrons, completely in accord with the molecular orbital picture. These experimental results do not show any sp³ hybrids at all.

Extending the results of Figure 3.3 and Structure **3.2** one can see that in general, if there are N pairs of electrons in N bonding orbitals in a molecule, then these may be localized to give N two-center, two-electron bonds. But what happens when there are fewer pairs of electrons in bonding orbitals than our chemical prejudice wants to make 'bonds'?

There are two choices. Either the delocalized picture has to be used or 'bond orbitals' have to be constructed which are spread over more than two atoms. The circle conventionally drawn inside the hexagon of benzene is an example of the first type. Here there are only three π-bonding pairs of electrons (Structure **3.5**), but six C–C close-contacts. Interestingly, in benzene one invariably uses the localized model to view the C–C and C–H σ manifolds, but the delocalized model (out of necessity) to view the π manifold. (Alternatively one can write a set of resonance structures.) So here the two models are used at the same time in the same molecule.

The localized picture has many other uses. It's behind the valence bond model of chemical bonding that may be used in a quantitative way to study chemical problems in the same way as molecular orbital theory via the discussion centered around the result of equation (3.7) or in the qualitative way that is so useful in all areas of chemistry. One

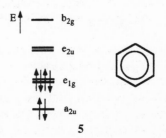

5

particularly nice application of the localized orbitals is as a basis for a delocalized picture. It includes a study of the σ manifolds of molecules and solids. Structure **3.6** shows a pair of atoms (as in a saturated hydrocarbon) connected by some sort of hybrid orbital. Let's assume that these are sp³ hybrids. We have shown two interaction integrals in the diagram, one connecting hybrids on adjacent atoms and one connecting hybrids on the same atom (on-site interaction). That there is an interaction between hybrids on the same atom often comes as a surprise but you have already noted that these localized hybrids are not orthogonal. However, let's show how large it is using a simple model. First we write the wavefunctions for two sp³ hybrids, $\xi_{1,2}$ on the same atom as

$$\xi_1 = (1/2)(\phi_s + \phi_x + \phi_y + \phi_z)$$
$$\xi_2 = (1/2)(\phi_s - \phi_x + \phi_y - \phi_z) \tag{3.9}$$

where the ϕ_i represent the s, p_x, p_y and p_z atomic orbitals. Now the interaction integral linking the two is

$$\beta_1 = (1/4)\langle(\phi_s + \phi_x + \phi_y + \phi_z)|\mathscr{H}^{\mathrm{eff}}|(\phi_s - \phi_x + \phi_y - \phi_z)\rangle \tag{3.10}$$

where $\mathscr{H}^{\mathrm{eff}}$ is the effective Hamiltonian for the problem. The result is

$$\beta_1 = (1/4)(e_s - e_p) \tag{3.11}$$

namely a stabilization equal to one quarter of the separation of the atomic s and p levels. For carbon these atomic levels lie at -19.4 eV and -10.7 eV leading to a value of β_1 of 2.2 eV, in diamond or in cyclohexane for example. For the polysilanes the comparable figure is 1.8 eV from the values of the atomic orbital energies. A smaller value is determined spectroscopically as we see below. The magnitude of β_2 depends on the internuclear separation.

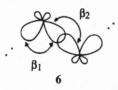

6

Such a simple model leads to a useful way[10] to view the origin of the band gap in diamond (Figure 3.5). β_2 gives rise to localized bonding and antibonding levels between all bonded pairs of atoms. The width of the energy bands comes from the on-site interactions. This process may be done more rigorously to generate the band structure itself.

Can we use this type of approach in molecules and study the variation in HOMO-LUMO gap in, for example, hydrocarbons of different topologies?

Yes, but we have to be careful and make sure it's a delocalized picture which results. The HOMO-LUMO gap is not a collective property. As an example we can organize[11] the set of ionization energies associated with the σ-bonds in the series of permethylated silanes of Structure **3.7**. Just as in the model described for methane earlier, if each σ-bond were truly localized between the corresponding pair of silicon atoms then not only would one peak be seen in the photoelectron spectrum of each molecule, but this would be at the same

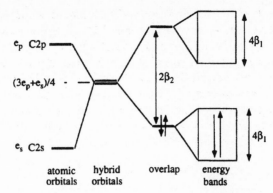

Figure 3.5
A way to derive the band structure of diamond. β_2 gives rise to localized bonding and antibonding levels between all bonded pairs of atoms. The width of the energy bands comes from the on-site interactions (β_1). Both valence and conduction bands are located asymmetrically about the bonding and antibonding hybrid energies. In each case the bottom of the band lies $3\beta_1$ below these energies. (For further discussion, see: Pettifor, D. G., Bonding and Structure of Molecules and Solids, Oxford University Press (1995).)

position for all members of the set, assuming equal Si−Si distances. Several ionizations however, are observed. First let's assume that the two electrons in an isolated Si−Si bond lie in an orbital with an energy of α_{Si-Si}. (This will have an energy related to β_2 above.) Adjacent localized bonds interact with each other *via* a term β_1 (Structure **3.6**). We can set up the secular determinants (we showed how to do this in general in Chapter 2) for all of the molecules shown in Structure **3.7** and express the occupied orbital energies in the form $E = \alpha_{Si-Si} + x\beta_1$. That for the cyclic $(SiH_2)_5$ molecule is shown in equation (3.12). We have dropped the subscripts for notational convenience. It is *topologically* identical to that for the $p\pi$ orbitals of cyclopentadienyl.

$$\begin{vmatrix} \alpha - E & \beta & 0 & 0 & \beta \\ \beta & \alpha - E & \beta & 0 & 0 \\ 0 & \beta & \alpha - E & \beta & 0 \\ 0 & 0 & \beta & \alpha - E & \beta \\ \beta & 0 & 0 & \beta & \alpha - E \end{vmatrix} = 0 \qquad (3.12)$$

Solution of the equation gives the energies of the σ bonding orbitals for the molecule. Manipulation of the whole set of data of Structure **3.7** in this way leads to the plot of Figure 3.6[11]. One can readily determine the two parameters of the model; $\alpha_{Si-Si} = 8.7$ eV

H_3Si-SiH_3	H_3Si-SiH_2-SiH_3	H_3Si-SiH_2-SiH_2-SiH_3
8.69eV	9.19	8.00
	8.22	8.78
		9.38

Si-$(SiH_3)_4$	cyclic $(SiH_2)_5$	cyclic $(SiH_2)_6$
8.26eV	7.80	7.75
	8.90	8.16
		9.18

7

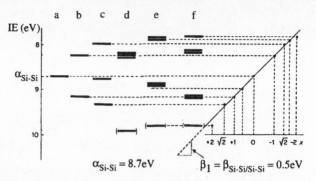

Figure 3.6

Observed photoemission energies for some polysilanes and the relative values predicted using the method described in the text. Plain solid lines correspond to the observed peaks in the photoelectron spectra of these molecules, their degeneracy guided by what should be expected from consideration of the theory. Solid lines in parentheses correspond to unobserved peaks. The scale of the diagram is set by the predictions of the theory (dashed lines). At the bottom right are the calculated energies in the form $E = \alpha_{Si-Si} + x\beta_1$ ($\beta_1 = \beta_{Si-Si/Si-Si}$). The best fit to the data is found for $\alpha_{Si-Si} = 8.7$ eV and $\beta_{Si-Si/Si-Si} = 0.5$ eV. (Adapted from reference 11. The letters a–f refer to the molecules of Structure 3.7.)

and $\beta_1 = \beta_{Si-Si/Si-Si} = 0.5$ eV. (For comparison the analogous value, $\beta_{C-C\pi/C-C\pi}$, for conjugated polyenes is about 0.6 eV from similar spectral data.)

But hybrids turn up in molecular orbital theory too.

There are in fact two different meanings of the term hybrid in common usage and we should be careful to distinguish between them. The first we have discussed at length here. These hybrids are described by a wavefunction with a directional character whose composition is set by chemical prejudice in the simplest picture. The sp³ orbitals in methane are an obvious example. Alternatively they are constructed using some localization procedure. The second is the automatic mixing of orbitals of the same symmetry in the molecular orbital description. Look at the orbitals of methylene in Structure **3.3**. One of the frontier orbitals is a pure carbon 2p orbital (b₂) by symmetry, the other (a₁) is a mixture, or hybrid, of carbon 2s and 2p. The ratio of the mixing coefficients is not integral (as in the sp³ hybrid of the localized approach) but is determined in a relatively complex way by the overlap integrals between carbon 2s and 2p with the hydrogen 1s orbitals, and the relative energies of all three atomic orbitals.

References

1. Woodward, R. B. and Hoffmann, R., *The Conservation of Orbital Symmetry*, Academic Press (1970).
2. See almost any introductory organic text.
3. See e.g., Cotton, F. A., *Chemical Applications of Group Theory*, 3rd Edition, John Wiley & Sons (1990).
4. Lennard-Jones, J. E., *Proc. Roy. Soc.*, **A198**, 1 (1949).
5. For example, Potts, A. W. and Price, W. C., *Proc. Roy. Soc., Lond.*, **A326**, 165 (1972).
6. Bader, R. F. W., *Atoms in Molecules*, Oxford (1994).
7. Savin, A., Becke, A. D., Flad, J., Nesper, R., Preuss, H. and von Schnering, H. G., *Angew. Chem., Int. Ed.*, **30**, 409 (1991).

8. Dewar, M. J. S., *Theory of Molecular Orbitals*, McGraw-Hill (1969).
9. Clark, S. A. C., Reddish, T. J., Brion C. E., Davidson, E. R. and Frey, R., *Chem. Phys.*, **143**, 1 (1990).
10. Bullett, D. W., *Solid State Phys.*, **35**, 129 (1980).
11. Bock, H., *Angew. Chem. Int. Ed.*, **28**, 1627 (1989).

General References and Further Reading

Atkins, P. A., *Quanta, A Handbook of Concepts*, Oxford University Press (1991).
Boys, S. F., *Rev. Mod. Phys.*, **32**, 296 (1960).
DeKock, R. and Gray, H., *Chemical Structure and Bonding*, Benjamin/Cummings (1980).
Dewar, M. J. S. and Dougherty, R. C., *The PMO Theory of Organic Chemistry*, Plenum (1975).
Edmiston, C. and Ruedenberg, K., *Rev. Mod. Phys.*, **35**, 457 (1963).
Edmiston, C. and Ruedenberg, K., *J. Chem. Phys.*, **43**, S97 (1965).
Jørgensen, C. K., *Orbitals in Atoms and Molecules*, Academic Press (1962).
Kutzelnigg, W., *Angew. Chem.*, **35**, 573 (1996).
McGlynn, S. P., Vanquickenborne, L. G., Kinoshita, M. and Carroll, G. G., *Introduction to Applied Quantum Chemistry*, Holt, Rheinhart and Winston (1972).
McWeeney, R., *Coulson's Valence*, Oxford (1979).
Murrell, J. N., Kettle S. F. A. and Tedder, J. M., *The Chemical Bond*, Second Edition, John Wiley & Sons (1978).
Pilar, F., *Elementary Quantum Chemistry*, McGraw-Hill (1968).

4 Why are the Elements of the First Short Period, Li–Ne (and the First Row Transition Metals) so Special?

The lighter elements of both the main group (Li–Ne) and transition metal series (Sc–Ni) receive perhaps the most attention from chemists. In fact the types of compounds they form and their reactivity are generally considered to represent the norm, and the behavior of the heavier elements are often considered as being exceptions. This attitude is somewhat understandable given the major rôle played by carbon-containing compounds, in organic chemistry and the biological world. But surely it is the properties of these first row elements, boron to fluorine that are unusual. CO_2 is found as a gas or molecular solid where the triatomic unit is preserved but all of the other Group 14 AO_2 systems are found as extended solid-state arrays. The same comment applies to the structure of elemental nitrogen and those of the other Group 15 elements. The usual explanation for these striking differences centers on the relative strength of σ and π bonding. Thus the bond energies associated with the one σ and two π bonds in the X_2 molecule are larger for nitrogen than the three single bonds found for its congeners in the solid-state structures of black phosphorus or arsenic or their P_4 or As_4 analogs (Figure 4.1). When comparing these parameters down a group, π bonding is regarded as being much less effective beyond the first row. One should be careful in comparing internuclear distances between elements of different Z, but Table 4.1 shows single bond distances for the group 14 elements. It is interesting to see that the major change takes place between carbon and silicon, with much smaller changes thereon. Thus one might wonder

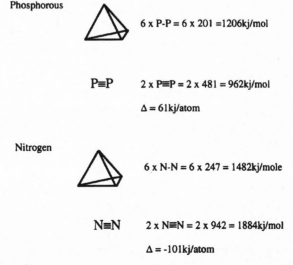

Phosphorous

6 x P-P = 6 x 201 =1206kj/mol

P≡P   2 x P≡P = 2 x 481 = 962kj/mol

Δ = 61kj/atom

Nitrogen

6 x N-N = 6 x 247 = 1482kj/mole

N≡N   2 x N≡N = 2 x 942 = 1884kj/mol

Δ = -101kj/atom

Figure 4.1
Some single and multiple bond energy comparisons for the Group 15 elements. We have used typical values but note that the strain energy in the P_4 tetrahedron as measured by its energy relative to black phosphorus, the more stable allotrope, is quite large. Δ is the energy difference between the two structures

Table 4.1 Single bond distances
in the group 14 elements

Element	Single bond distance, d (Å)
Carbon	1.54
Silicon	2.34
Germanium	2.44
Tin	2.80
Lead	2.88

whether the difference in behavior between the first row and heavier elements is determined largely by the much shorter internuclear separations in the former than in the latter. Short internuclear distances stabilize multiple bonds since pπ–pπ overlap drops off much faster with increasing distance than does pσ–pσ overlap. Thus, as indicated for nitrogen and phosphorus, molecules rather than extended solid-state arrays are often found for the first row elements both as compounds and in their elemental state. Certainly molecular dimensions bear this out, but is what is behind this striking 'first-row anomaly'?

You are, in fact very close to the real state of affairs here. Your comments however, bring up the very old question of 'atomic size' and how you estimate it. Sir Lawrence Bragg developed his first ideas in 1920 and many of them[1] were incorporated into a famous study by Linus Pauling[2]. They were able to assemble a size scale based on a partition of the interatomic distance between two atoms in a crystal. Such collections[3] of radii are often useful in estimating the dimensions of molecules and solids. However, the very idea of a unique radius is often a dubious one, especially so given the probabilistic nature of the behavior of an electron. Recall, there is a non-zero, but very small probability of the electron being far from the nucleus. Interatomic distances change with coordination number too; those for six-, larger in general than those for four-coordination.

A very useful model with which to study this problem is accessible using the idea of the pseudopotential[4–7]. It describes the potential felt by an outer, valence electron in an atom. The core electrons are assumed to be frozen, and so one studies the behavior of the valence electrons as determined by their interaction with the nucleus and this collection of inner electrons. There are just two terms, the Coulombic attraction of the electron with the nucleus, but screened because of the presence of the electronic core that lies between the two, and the repulsion between the core and valence electrons. The two nearly cancel each other so that the pseudopotential is rather weak when compared both to the raw Coulombic attraction and that screened by the core. Recall that the ionization potential for hydrogen-like atoms is given by Z^2R where R is the Rydberg constant equal to 13.6 eV. So without such electron–electron interactions, the first ionization potential of carbon would be $Z^2 \times 13.6 = 6^2 \times 13.6 = 489.6$ eV rather than the observed 11.26 eV. Figure 4.2 depicts a typical pseudopotential that shows the balance between the two terms. It allows a unique theoretical definition of a radius, the point where there is a radial maximum in the pseudowavefunction. This occurs at the point where the repulsive and attractive forces exactly balance, the crossing point of the pseudopotential (R_c).

I can readily see the origin of the attractive part of the potential, but what about the repulsive part?

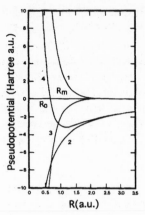

Figure 4.2
A typical pseudopotential: 1 = Coulomb potential; 2 = screened Coulomb potential; 3 = Pauli repulsion; 4 = pseudopotential

This comes about through the Pauli repulsion associated with the energy penalty as two electrons with the same l quantum number, but different n, enter the same region of space. The pseudowavefunction is orthogonal to all orbitals with different l by virtue of their angular behavior, but is not orthogonal to the core orbitals of the same l since the radial part has not been chosen to be that way. It is a pseudowavefunction after all and does not come from solution of the Schrödinger wave equation where such orthogonality is required. Thus we can imagine that these valence-core interactions lead to an energetic repulsion, in exactly the same way as the repulsion found for two helium atoms in Chapter 15. We will describe there why this is called a 'Pauli' repulsion. Thus the repulsion experienced by the 3p electrons in sulfur comes from the interaction with the 2p electrons. Such a repulsion, combined with the screened Coulomb attraction, sets the value of the pseudopotential radius and thus the 'size' of the atom. So there are different values of R_c for valence s, p and d orbitals, the pseudopotential radii r_s, r_p, r_d. But a rather special situation applies to the 2p orbitals of nitrogen and oxygen for example. Here there is no 1p orbital to provide such a repulsion and the electron–nucleus distance is set by a similar but more complex picture than that described for the atom in Chapter 1. The result is a much smaller atomic 'size' for the elements Li–Ne than that found for the heavier congeners.

This is really a rather simple result. First row atoms are much 'smaller' than their heavier congeners. The shorter distances, as I pointed out, will magnify the importance of multiple bonds formed by first-row elements.

There are some other results that come from pseudopotential theory that also give us more insights into this interesting problem. One approach to understanding the structures of solids gives[5] the total energy in a particularly simple form as

$$E_T = E_\Omega + \tfrac{1}{2}\sum_{ij} \Phi(R_{ij}) \qquad i \neq j \tag{4.1}$$

The Φ are effective two-body potentials that describe the interaction between atoms i and j, which lie a distance R_{ij} apart. The term E_Ω is a volume dependent term but one that is structure independent. So the parameters $\Phi(R_{ij})$ are not general interaction potentials, but are those for the case where the atoms are arranged in a box of constant volume. Figure 4.3 shows[8] a set of potentials for some of the elements of the second row. One can identify three main regions. At small R there is a strong repulsive contribution as expected. At very large distances the potential is an oscillating one gradually dropping off in amplitude. This part will not concern us. The behavior at intermediate distances is most important. In our discussions of structural possibilities we have to realize that we work from equation (4.1) at constant volume, so that if some distances are shortened, others have to lengthen as a consequence. Obviously the system will search out that geometry where the largest number of neighbors fall into the wells of the two-body potential. It is here where the total energy is minimized.

> So, if some atom pairs fall on a peak then the energy can be reduced by moving some atoms closer together and some atoms further away from their neighbors so they fall into a minimum in the pair potential.

Yes. One point on these figures is particularly important. This is the value of R expected if the solid were close-packed (D_{cp}) and is indicated by a circle in Figure 4.3. For Na, Mg

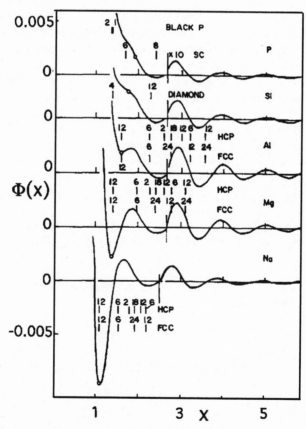

Figure 4.3
A set of potentials ($\Phi(R_{ij})$) for some of the elements of the second row (x is a reduced parameter that contains R)

and Al the value of D_{cp} is such that the nearest neighbors lie in the first minimum (R_{min}) in $\Phi(R)$. Thus close-packed structures are expected and are indeed found for these elements (Γ, the coordination number $= 12$). For Si and P, however, this is not the case. Notice that R_{min} lies to larger R than D_{cp}. As a result, more open structures are expected with a smaller number of closer neighbors (moving uphill in energy on the $\Phi(R)$ plot) and, as you noted, a number of more distant neighbors moving downhill. For silicon this works in the following way. Four neighbors move inward (energetically penalizing) leading to a smaller first coordination number ($\Gamma = 4$). However, there are twelve second-nearest neighbors lying at a longer distance but at a point of energetic stability, R_{min}. A similar result is found for phosphorus. The chemical trends on moving from left to right across the Periodic Table, increasingly open structures, are mimicked nicely by the form of the $\Phi(R_{ij})$. The theory gets right too, the preference for the hcp or fcc structure for the first three elements. There are more neighbors for Na and Mg in the second attractive well for the hcp structure than for fcc, and so the hcp structure is found. However, for Al the fifth nearest neighbors would lie at the maximum of the second repulsive wiggle in the hcp structure and the fcc arrangement is the one predicted and observed.

Presumably $\Phi(R_{ij})$ is quite different for the first row elements.

This is indeed where we are leading. A suggested $\Phi(R_{ij})$ function[9] is shown in Figure 4.4 for a typical first row element. It does have the same shape as that found for Na or Mg (Figure 4.3), but the value of D_{cp} is found at a larger value of R than R_{min}. This implies again that some atoms will move in to closer distances but others will move further away from the close-packed arrangement to give an open structure. This is analogous to the situation for silicon and phosphorous but is much more exaggerated for the first row case. There is a rough relationship that links coordination number (Γ) with these two parameters (equation (4.2)).

$$R_{min}/D_{cp} = [(\Gamma + 1)/13]^{1/3} \qquad (4.2)$$

Using the values from Figure 4.4 a figure of $\Gamma = 12$ is found for Na (in accord with its close-packed structure) but 0 for the first row element. Obviously 0 is too small but it does show that it is the form of the pseudopotential, without a strong repulsive contribution at low R that leads to small coordination numbers for these first row elements.

The origin of the result is, as noted earlier, the lack of a filled 1p core to provide a repulsion (Figure 4.2) of the 2p valence electrons.

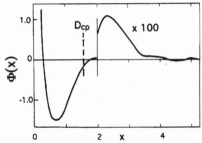

Figure 4.4
A suggested $\Phi(R_{ij})$ function for a typical first row element (x is a reduced parameter that contains R)

[handwritten annotations:
$Cr^6 = 273 eV.$
$Mo^{6+} = 226 eV.$

Gp 13, p2 =B Al Ga In Tl
14 p2 =C Si Ge Sn Pb
15 " =N P As Sb Bi

p2 p3 p4 p5 p6*]

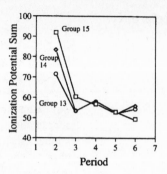

Figure 4.5

The variation in the sum of the first three ionization energies for three series of main group elements (units eV)

Does the small size of these first row atoms show up in high ionization potentials too?

Yes. Figure 4.5 shows the variation in the sum of the first three ionization energies for some of the main group elements. The smaller 'size' results in the valence electrons lying closer to the nucleus and thus being more strongly bound in electrostatic terms. A similar trend is found in their electronegativities too (see Chapter 2).

And the same type of effect is found for the first row transition metal series since the absence of a 2d orbital leads to contracted 3d orbitals.

Yes, but here though, the effect on the internuclear separations is not quite as marked as for the main group situation, since the 4s and 4p orbitals on the metal are also important in chemical bonding to the ligands. There are many observations, that may be associated with this contraction. The interesting observation concerning metal–insulator transitions in materials containing first row transition elements (the cuprate superconductors probably now being the best known) are among them. The interplay of high- and low-spin compounds and the solid state analog, ferromagnetism, is a phenomenon by and large found only in the first row metals. First row compounds are often colored but their analogs in the 4d and 5d series frequently are not. Both of these observations are explicable given the nature of the contracted 3d orbitals and their smaller interaction with the ligand orbitals, compared with analogous interactions involving 4d and 5d orbitals. The result is smaller d orbital splittings in complexes containing first-row transition metals.

These effects are found for elements at the top of the Periodic Table. Does anything interesting happen at the bottom?

Actually, yes. For very heavy atoms, relativistic effects come into play in a major way since electrons that get close to heavy nuclei are strongly accelerated. Since only s orbitals have a non-zero value of the wavefunction at the nucleus, they are affected the most. The result is a contraction of these orbitals and especially so at the bottom of the Periodic Table. Many interesting features[10,11] result. The yellow color of elemental gold is one of them. Bond lengths are also shorter for the heavier elements (Au(I) is almost 0.1 Å smaller than Ag(I) when bound to phosphine) as a result of such a contraction. Frequently bond strengths are higher too.

References

1. Bragg, W. L., *Phil Mag.*, **40**, 169 (1920).
2. Pauling, L., *J. Amer. Chem. Soc.*, **51**, 1010 (1929).
3. Shannon, R. D., *Acta Cryst.*, **A32**, 751 (1976); Shannon, R. D., *Structure and Bonding in Crystals*, Vol. 2, O'Keeffe, M. and Navrotsky, A. (eds), Academic Press (1981).
4. Zunger, A., *Structure and Bonding in Crystals*, Vol. 1, O'Keeffe, M. and Navrotsky, A. (eds), Academic Press (1981).
5. Heine, V., *Solid St. Phys.*, **24**, 1 (1970).
6. Cohen, M. L. and Heine, V., *Solid St. Phys.*, **24**, 37 (1970).
7. Kutzelnigg, W., *Angew. Chem., Int. Ed.*, **23**, 272 (1984).
8. Alonso, J. A. and March, N. H., *Electrons in Metals and Alloys*, Academic Press (1989).
9. Hafner, J. and Heine, V., *J. Phys.*, **F13**, 2479 (1983).
10. Pyykkö, P., *Chem. Rev.*, **88**, 563 (1988).
11. Bersuker, I. B., *Electronic Structure and Properties of Transition Metal Compounds*, John Wiley & Sons (1996).

General References and Further Reading

Alonso, J. A. and March, N. H., *Electrons in Metals and Alloys*, Academic Press (1989).
Cohen, M. L., *Structure and Bonding in Crystals*, Vol. 1, O'Keeffe, M. and Navrotsky, A. (eds), Academic Press (1981).

5 Why does the Molecular Orbital Model often give the Wrong Products of Dissociation?

If I take the molecular orbital diagram for a molecule such as that of HF in Figure 5.1 and use the fact that overlap decreases with increasing distance I can draw the diagram of Figure 5.2. It shows that on stretching the HF bond the molecule dissociates to $H^+ + F^-$, i.e., the two electrons in the σ-bonding orbital, largely located on fluorine in the molecule, end up completely localized on the fluorine atom (i.e., F^-) in the products. But this cannot be correct. The energy of $H^+ + F^-$ lies higher than that of $H \cdot + F \cdot$ by 10.3 eV, the sum of the ionization energy of H (13.6 eV) and the electron affinity of fluorine (3.3 eV). This seems to be a problem for the molecular orbital model.

This is indeed true. There are problems too in calculating the dissociation energies of molecules, even when we know what the lowest energy products are. For example, from an excellent Hartree–Fock calculation the H–F bond in HF is calculated[1] to be 4.11 eV, to be compared with the 6.08 eV found from experiment. Calculation of bond energies is in general quite difficult. This problem may be tackled in a variety of ways but let's see first where it comes from. We will use the H_2 molecule as a simple example. Although in Chapter 3 we stressed the importance of an antisymmetrized wavefunction, we can retain the essence of the quantum mechanics here by writing a simpler wavefunction to describe the situation in the H_2 molecule which contains two electrons in the bonding orbital, $\psi_b = 1/\sqrt{2}(\phi_1 + \phi_2)$, and no electrons in the antibonding orbital, $\psi_a = 1/\sqrt{2}(\phi_1 - \phi_2)$, where the $\phi_{1,2}$ are atomic H 1s orbitals. (We have ignored overlap in the normalization process for simplicity here.) Equation (5.1) is a Mulliken–Hund wavefunction[2] and

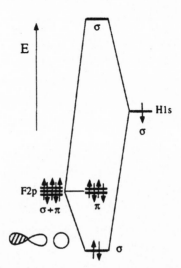

Figure 5.1
The molecular orbital diagram for HF. The form of the bonding σ orbital is shown

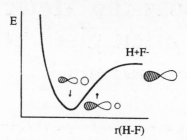

Figure 5.2
Variation in energy and in character of the lowest occupied orbital of Figure 5.1 on stretching the HF bond

contains two types of terms, those, such as $\phi_1(1)\phi_1(2)$ where the two electrons reside on the same atom (ionic terms) and those where they lie on different atoms (covalent terms), e.g., $\phi_1(1)\phi_2(2)$:

$$\begin{aligned}\Psi_{MH} &= \psi_b(1)\psi_b(2) \\ &= \tfrac{1}{2}(\phi_1 + \phi_2)(1)(\phi_1 + \phi_2)(2) \\ &= \tfrac{1}{2}[\phi_1(1)\phi_1(2) + \phi_2(1)\phi_2(2) + \phi_1(1)\phi_2(2) + \phi_2(1)\phi_1(2)]\end{aligned} \qquad (5.1)$$

You can evaluate the total electronic energy for this wavefunction within the Hückel approximation as

$$E = 2(\alpha + \beta) + U/2 \qquad (5.2)$$

where we have introduced the electron–electron repulsion term $U = \langle \phi_1(1)\phi_1(2) \,|\, r_{12}^{-1} \,|\, \phi_1(1)\phi_1(2)\rangle$.

I can see there must be a term containing U in the energy since there is a non-zero probability from the form of the wavefunction (equation (5.1)) of finding both electrons on the same atom at the same time.

In a similar fashion we can write a wavefunction for the situation where there are two electrons in the antibonding orbital.

$$\Psi'_{MH} = \psi_a(1)\psi_a(2) \qquad (5.3)$$

You can show it has an energy $E = 2(\alpha - \beta) + U/2$.

Now let's see what happens to the bonding state of equation (5.1) as we move the two hydrogen atoms apart. The ionic terms where the two electrons reside on the same atom are now less appropriate and will in fact have zero weight at an infinite separation if we insist on the presence of two hydrogen atoms in their ground state (H·) rather than H$^+$ + H$^-$. Clearly at this point the wavefunction of equation (5.1) is now inappropriate and a (Heitler–London)[3] wavefunction should be written as

$$\Psi_{HL} = 1/\sqrt{2}[\phi_1(1)\phi_2(2) + \phi_1(2)\phi_2(1)] \qquad (5.4)$$

Its energy is just $E = 2\alpha$. In crude terms one can compare this with the energy of the Mulliken–Hund wavefunction, $2(\alpha + \beta) + U/2$. The energy difference between the two $(2\beta - U/2)$ represents a balance between energetic terms involving one electron (β) and those involving two electrons (U).

So, on the simplest model the bond energy is a trade-off between the two.

Yes. Returning to the difference between the two wavefunctions it is clear to see that the electrons cannot in general be allowed to move completely independently of each other. In the Mulliken–Hund picture the electrons are completely uncorrelated, (i.e., completely unrestricted in their movement) and in the Heitler–London, completely correlated, i.e., are completely restricted in their movement. One can expect that a situation between the two will frequently, perhaps invariably, be found. There are several ways to take this into account. The most common approach in molecular chemistry[4,5] to tackle this problem of electron correlation uses a technique known as configuration interaction (CI). Let's write a new wavefunction, Ψ_{CI}, which is a linear combination of the two Mulliken–Hund wavefunctions, Ψ_{MH}, Ψ'_{MH} we derived earlier:

$$\Psi_{CI} = \Psi_{MH} + \lambda \Psi'_{MH}$$
$$= \tfrac{1}{2}[(1 + \lambda)[\phi_1(1)\phi_1(2) + \phi_2(1)\phi_2(2)] + (1 - \lambda)[\phi_1(1)\phi_2(2) + \phi_2(1)\phi_1(2)]] \quad (5.5)$$

The mixing parameter (λ) is in practice determined numerically, but you can see that if $\lambda < 0$, mixing in of some Ψ'_{MH} into Ψ_{MH} increases the weight of the ionic terms such as $\phi_1(1)\phi_2(2)$ and reduces the weight of the covalent terms such as $\phi_1(1)\phi_1(2)$. Thus the configuration interaction scheme nicely takes into account the electron correlation. In larger molecules though, as you can imagine, this process becomes quite complex. Of course all of these calculations are performed numerically. Often though, the equilibrium geometry found from a Hartree–Fock calculation (one without the addition of CI) will be very similar to that obtained from the calculation where CI is included.

In general the curve of Figure 5.2 should be modified as in Figure 5.3 for the general case, to overcome the problem of incorrect dissociation products. As the bond is stretched, more of the covalent terms are mixed into the ground state as the energy separation between the two states becomes smaller and the result is an avoided crossing of the two states along the abscissa.

Does this mean that we always need to include electron correlation in this way when studying molecules and solids with different coordination numbers, i.e., where bonds have been broken?

We have mentioned briefly the most popular molecular method (CI) above. For solids, the present method of choice is to use Density Functional Theory. This is an approach that writes the exchange and correlation parts of the energy as a function of the electron density. This is a technique suggested by John Slater[6] and, other than just mentioning it, we

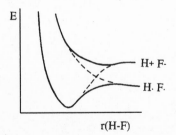

Figure 5.3
The avoided crossing along the abscissa of the two states that correspond to $H^+ + F^-$ and $H \cdot + F \cdot$ at long distances

won't discuss it further. It is a vital part of the LMTO and LAPW electronic structure methods of the solid-state physicist and more recently has been used in calculations on molecules. However, there is another much simpler method that has proven quite useful in the calculation of the relative energetics of both molecules and solids. As we show in Chapter 13 the concept of the moments of the electronic density of states is a very useful one[7,8]. The nth moment of the collection of energy levels $\{E_i\}$ is simply given by the expression

$$\mu_n = \sum_i E_i^n \tag{5.6}$$

There is in principle an infinite set of moments, but here we will just consider the importance of the second moment. We will not discuss here the theory behind the result[8,9] at all but just state that in terms of deciding energetically between structural alternatives, good agreement between theory and experiment is found using simple Hückel theory if the relative energy is evaluated from calculations where the second moments of the electronic densities of states are set equal. You can take a very simple example, the triangular and linear structures of Structure 5.1 for the H_3 molecule and its ions.

H   H———H———H
H———H
1

Yes, to show this I can draw out in Figure 5.4 the Hückel energy levels for the two geometries using two different Hückel β values, since the molecules have different internuclear separations. There is no reason to expect that the two βs will be the same. The figure also shows the energy levels of the linear molecule in terms of the β values of the triangular species which I can generate by setting the second moments of the two systems equal in the following way:

$$3\alpha^2 + (\sqrt{2}\beta')^2 + 0^2 + (\sqrt{2}\beta')^2 = 3\alpha^2 + 2\beta^2 + (2\beta)^2$$

so

$$4\beta'^2 = 6\beta^2$$
$$\beta' = \sqrt{3/2}\beta \tag{5.7}$$

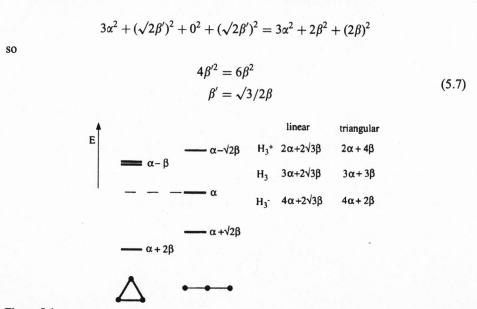

Figure 5.4
Hückel levels and energies for the open and closed geometries of H_3

So, from your results H_3^+ should be triangular (which indeed it is) but H_3 and H_3^- linear. The latter species are unknown (see our discussion of Chapter 12) but the Jahn–Teller theorem does tell us they should be unstable in the triangular geometry.

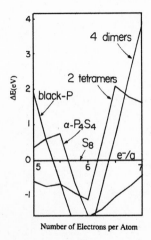

$$\alpha\text{-}P_4S_4 \xrightarrow{4e^-} S_8 \xrightarrow{4e^-} 2\, S_4^{-2} \xrightarrow{4e^-} 4\, S_2^{-2}$$

2

A more complex example is shown in Figure 5.5, which compares[8] the relative energies calculated for the molecules shown in Structure **5.2** obtained using this idea of second-moment scaling. The structure of black phosphorus (Figure 4.1) contains three-coordinate atoms, and in accord with the simple ideas of the Lewis electron-pair bond, the coordination numbers in the other systems decrease as the electron count increases. The observed structures for a given electron count are nicely reproduced by the calculation.

What happens without second moment scaling?

The structure with the highest coordination number is usually computed to be the most stable.

Similar comments apply to the examples shown in Figure 5.6. The correct structure is found[10] by calculation each time. The hcp structure is indeed found for lithium and beryllium and the unusual rhombohedral R-12 structure is found for boron. The correct energetic separation of graphite and diamond is found by calculation. Without second moment scaling, the diamond structure is generally computed to be more stable by simple calculation.

Figure 5.5
*Computed relative energies of the molecules of Structure **5.2** using second-moment scaling*

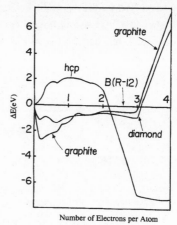

Number of Electrons per Atom

Figure 5.6
Computed relative energies of the elemental solids of the first long period (Li–C) using second-moment scaling

References

1. Demuynk, J., Veillard, A. and Wahlgren, U., *J. Amer. Chem. Soc.*, **95**, 5563 (1973).
2. Mulliken, R. S., *22nd Annual Congress of Pure and Applied Chemistry*; *Plenary Lectures*, Butterworth (1970).
3. Heitler, W. and London, F., *Z. Phys.*, **44**, 455 (1927).
4. Hehre, W. J., Radom, L., Schleyer, P. v R. and Pople, J. A., *Ab Initio Molecular Orbital Theory*, John Wiley & Sons (1986); Pople, J. A., Binkley, J. S. and Seeger, R., *Int. J. Quantum Chem.*, **S10**, 1 (1976).
5. Bersuker, I. B., *Electronic Structure and Properties of Transition Metal Compounds*, John Wiley & Sons (1996).
6. Slater, J. C., *Adv. Quantum Chem.*, **6**, 1 (1972); Slater, J. C. and Johnson, K. H., *Phys. Rev.*, **B5**, 844 (1972).
7. Burdett, J. K., *Chemical Bonding in Solids*, Oxford University Press (1995).
8. Lee, S., *Accts. Chem. Res.*, **24**, 249 (1991).
9. Pettifor, D. G. and Podloucky, R., *J. Phys. C.*, **19**, 285 (1986).
10. Lee, S., Rousseau, R. and Wells, C., *Phys. Rev.*, **B46**, 12121 (1992).

General References and Further Reading

Berry, R. S., Rice, S. A. and Ross, J., *Physical Chemistry*, John Wiley & Sons (1963).
Kutzelnigg, W., *Angew. Chem., Int. Ed.*, **35**, 573 (1996).
Murrell, J. N., Kettle S. F. A. and Tedder, J. M., *The Chemical Bond*, 2nd Edition, John Wiley & Sons (1978).
Pople, J. A. and Beveridge, . ., *Approximate Molecular Orbital Theory*, McGraw-Hill (1970).
Richards, W. G. and Cooper, D. L., *Ab Initio Molecular Orbital Calculations for Chemists*, Second Edition, Oxford University Press (1982).
Schaefer, H. F., *Quantum Chemistry*; *The Development of Ab Initio Methods in Molecular Electronic Structure Theory*, Oxford University Press (1984).
McWeeney, R., *Methods of Molecular Quantum Mechanics*, Second Edition, Academic Press (1989).

6 How Important are d Orbitals in Main Group Chemistry?

In constructing hybrid orbitals, Linus Pauling used the valence s and p orbitals to construct sp^3 hybrids for tetrahedral carbon for example (Structure **6.1**). This model has been enormously successful in understanding the chemistry of octet compounds, especially organic ones. However, for molecules such as SF_6 where there are six ''bonds'' he had to invoke the use of the valence d orbitals to construct a set of six d^2sp^3 hybrid orbitals (Structure **6.2**) with which to make six two-center two-electron bonds. But just how important are d orbitals in main group chemistry?

1   2

For many years[1] valence d orbital involvement was used to 'explain' various unusual properties (structure for example) of main group molecules. Species such as these are often called hypervalent and the term 'expansion of the octet' was used for many years to describe this d orbital activity. We can put this in perspective however, by deriving molecular orbital diagrams generated with and without them. Figure 6.1 shows a molecular orbital diagram for the σ manifold of octahedral SF_6. As you can see, there are four bonding orbitals (from symmetry matches with the sulfur 3s and 3p orbitals). Thus the bond order of each S–F linkage is 2/3. Although we would expect the S–F distance in SF_6 to be a little longer than in SF_2, in fact they are very similar (1.56 Å and 1.59 Å respectively) showing that one cannot use bond lengths as a sole criterion of bond order, especially in molecules with different coordination number. The result of this orbital description is that we have to use the delocalized orbital picture (or a set of resonance structures) to describe the bonding here since six two-center two-electron bonds cannot be constructed. Notice though that there is a pair of ligand orbitals, of e_g symmetry that find no symmetry match with the sulfur orbitals and thus remain non-bonding. So, using the discussion of Chapter 7 the molecule is an octet compound with two ligand-located lone pairs. Therefore the octet rule is not really violated, there is no expansion of the octet and there are four bonding pairs of electrons.

So SF_6 is similar electronically to the $W(CO)(C_2H_2)_3$ molecule of Chapter 8. There, counting electrons formally leads to a total of twenty (two more than the stable electron count of eighteen for transition metal systems), but there is one ligand combination which by symmetry cannot interact with a central atom orbital. In the same way, SF_6 has an e_g pair that cannot interact, by symmetry with any of the central atom s, p orbitals too.

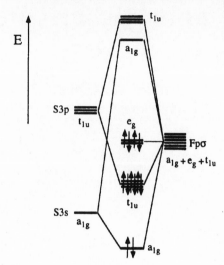

Figure 6.1
A molecular orbital diagram for the σ manifold of octahedral SF_6 with no 3d orbitals on sulfur

Exactly. But let's see what happens when we include the 3d orbitals on sulfur. Figure 6.2 shows a diagram, similar to Figure 6.1, but with the inclusion of 3d orbitals on the central atom. These transform as $e_g + t_{2g}$, and thus interaction with this hitherto non-bonding ligand e_g pair is allowed by symmetry. We show a relatively weak interaction in this figure. The sulfur 3d orbitals lie a long way energetically from the fluorine orbitals and so the interaction will not be very large. Thus the 3s/3p sulfur orbital picture is the one to use.

So the picture is quite straightforward. Sulfur 3d orbitals are really not very important. The Pauling picture that uses d^2sp^3 hybrids on the central atom to force the use of two center-two electron bonds is not correct. And it looks as though their historical usage is associated with a reluctance to write anything other than two-center two-electron bonds in molecules. The delocalized approach provides a perfectly acceptable picture.

Yes. These comments, which emphasis somewhat basic questions of bonding carry over to the picture extracted from numerical calculation. An extended Hückel calculation can be performed with and without sulfur 3d orbitals for SF_6 and the results are rather similar. There is indeed a small admixture of the higher energy orbitals into the occupied orbitals along the lines of Figure 6.2, but the general picture is little changed. Analogously although the details of the computed bond angles change by a small amount on inclusion of d orbitals, the general picture for the SF_4 molecule is well reproduced by the sp-only picture. In *ab initio* studies, the picture is similar. Here large basis sets may be used to alleviate the limitations of the LCAO approximation of equation (1.5). These will naturally include d functions on sulfur, and there will be some sulfur 3d character in the ground state wavefunction but these SCF calculations too show[3] a minimal rôle for the 3d orbitals and the basic chemical picture of an octet compound that uses sulfur sp orbitals with two lone pairs is still retained. The d^2sp^3 picture is not the best way to view these compounds.

Why then do these 'hypervalent' molecules form in the first place?

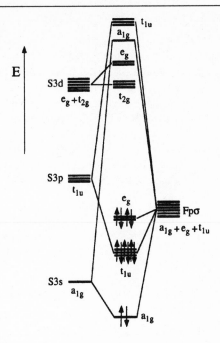

Figure 6.2
A molecular orbital diagram for the σ manifold of octahedral SF₆, but with the inclusion of 3d orbitals on sulfur

The question of their stability is a tricky one but no more so than in other parts of chemistry. Current thinking is that one of the important considerations is a steric one. Central atom–ligand distances in compounds involving first row atoms are quite short, but are much larger for their heavier congeners. These short distances lead to close non-bonded contacts and large repulsions between the ligands for compounds with central atom orbitals from the first row, but longer ones and smaller repulsions for their heavier analogs. On this basis, it is easier to coordinate six fluorine atoms around a sulfur atom than around an oxygen atom.

Is this then what's behind the general lack of 'hypervalent' molecules with a first row central atom?

It will certainly be a contributing factor. Since there are no 2d orbitals, Pauling's hybrid model has long been used to explain the paucity of simple compounds of 'five-valent carbon'. There are though quite a large number of molecules and solids containing first row atoms with coordination numbers greater than four. There are indeed a few species[2] that contain 'five-valent carbon' in a trigonal bipyramidal geometry. But even here d orbitals are not needed to understand the differences between the stability of CH_5^- and SiH_5^- as we show in Chapter 12. There are many molecules too of the type shown in Structures **8.28** and **8.29** that contain 'five- and six-valent carbon' although these have been commonly regarded as being 'electron deficient' (see Chapter 11) and so have not generally been regarded as being hypervalent. But they are. They are well described by a delocalized valence s/p model that does not involve d orbitals. Indeed there are an enormous number of boranes and their derivatives that contain five-coordinate boron, another first row element where there cannot be any d orbital involvement of the Pauling type. Their chemistry is described in Chapter 11 using the delocalized orbital model without the use of d orbitals. As time progresses such a delocalized viewpoint becomes less and less unusual. The use of the orbital approach gives quite a consistent picture.

Another place where d orbitals are often invoked is as acceptor orbitals (Structure **6.3**) in PR_3 ligands when coordinated to transition metals and in SiH_3 when stabilizing the planar geometry of ammonia as in Structure **6.4**. What is their rôle here?

3

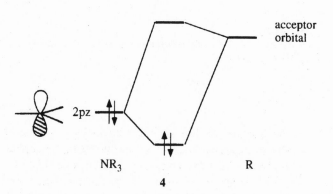

4

It is now quite well established that the orbitals responsible for both of these effects is a $\sigma*$ orbital involving the silicon or phosphorous and the three hydrogen atoms as in Structure **6.5**. Nitrogen does not behave similarly in ammonia since in this molecule the N−H distance is considerably shorter. The $\sigma*$ orbital then lies much higher in energy as a consequence and is much less important. As a result, NH_3 is not a good acceptor when coordinated to transition metals.

5

References

1. For an excellent discussion of this whole question from the viewpoint of an experimental chemist see: Gilheany, D. G., *Chem. Rev.*, **94**, 1339 (1994).
2. Martin, J. C., *Science*, **221**, 509 (1983).
3. Reed A. E. and Weinhold, F., **108**, 3586 (1986).

General References and Further Reading

Pauling, L., *The Nature of the Chemical Bond*, Third Edition, Cornell University Press (1960).

7 What's the Best Way of Looking at the Interaction of Transition Metals with their Ligands, and, What's Behind the Eighteen-electron Rule?

For many years the interaction of transition metals with the ligands attached to them was usually described by the Crystal Field Model. It was largely used to interpret electronic spectra, although it was able to provide an interesting electronic picture of the variation in the heats of formation of metal oxides, chalcogenides and halides. It was also able to tackle the variation in the rates of ligand loss from octahedral metal complexes across the first row transition metal series, and thus provide an electronic underpinning of the experimental observation of labile and inert complexes. Since the model is a purely ionic one it has a problem with the fact that the largest values of the e_g/t_{2g} splitting are found for neutral molecules such as CO and C_2H_4. We sometimes see today the term ligand field theory that appears to vary in its meaning. It's often used to describe an improved version of crystal field theory that includes some orbital interactions to take into account problems of the type found for CO. It is used too to describe an approach that looks very much like molecular orbital theory itself.

First we should state that there is no 'best' way to study molecules or solids. The theoretical model one uses in any area of chemistry should be linked to the type of problem under consideration. The Crystal Field Theory (CFT) is one way to readily construct the energy level splittings in octahedral and tetrahedral compounds. Parameterized properly it is a useful way to look at their spectra[1]. It certainly has had a long life. Its appeal is that it is simple to understand, although simplicity is not always associated with the correct result. The important thing to realize about the CFT as usually presented, is that it takes credit for results that are really symmetry based. Let's take the d orbital levels of a transition metal atom. In the gas phase (point group K_h) the five d levels transform as the d irreducible representation of this group. (Recall that the irreducible representations in this group are labeled s, p, d, f, ...) This representation is five-fold degenerate and so the five d orbitals have the same energy. In an octahedral field (point group O_h) in a crystal, or in an octahedral complex of the same symmetry, the highest degeneracy is three.

So by symmetry the d orbitals have to split into e_g and t_{2g} sets as in Structure **7.1**, separated in energy by Δ, although symmetry arguments as usual do not tell us which set lies lower in energy.

Of course, nothing at all has yet been said about chemical bonding, but the premise of the CFT is that the interactions are electrostatic ones, although any scheme that split the levels this way would work. Algebraically the splitting energies are obtained by applying, as a perturbation, the field associated with the six negative charges ($-ze$) arranged at the vertices of the octahedron with the ion core (charge Ze) at the center. The metal-ligand

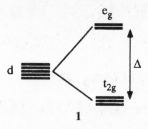

1

distance is a. The field contains a spherical part that raises the energy of all of the orbitals by $6Zze^2/a$, and a part that recognizes the specific ligand location. This second part has no effect on the energies of s and p orbitals but leads to a splitting of the d levels by $10Dq$ where $Dq = (1/6)(Zze^2/a^5)\langle r^4 \rangle$. $\langle r^4 \rangle$ is the expectation value of the fourth power of the electron–nucleus distance. One can simply see why the two d sets are inequivalent. The dz^2 and $x^2 - y^2$ orbitals point directly toward the negative charges (Structure **7.2**) and thus an electron in one of these orbitals experiences a larger electrostatic repulsion than when located in xz, yz or xy which point between the ligands. The center of gravity of the d levels is maintained (a result of the perturbation theory approach) so that three drop in energy by

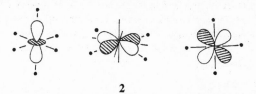

2

$2/5\Delta$ or $4Dq$ and two are raised in energy by $3/5\Delta$ or $6Dq$. One of the appealing features of the CFT is that it is easy to generate energy level splitting patterns for high symmetry structures, although for non-cubic groups (most of the systems that do not have tetrahedral or octahedral symmetry) the process is more complex[2]. However, this is a feature rarely discussed.

But most calculations performed today on transition metal complexes use molecular orbital methods, ranging from Hückel ideas through *ab initio* calculations.

Yes, the work of Roald Hoffmann in particular has shown the tremendous utility of such approaches in a wide variety of metal-containing systems[2]. The techniques are no different than those we use to look at methane, for example. The term 'ligand field theory' therefore, is one we should drop. There is however, a molecular orbital 'analog' of the CFT that is much more powerful and simpler to use. This is the Angular Overlap Model (AOM)[2,4].

Let's start by looking at the interactions between a pair of orbitals of different energy or electronegativity located on atoms M and X as in an MX_n system (Figure 7.1). (We did this in Chapter 2 too.) The two atoms have different electronegativities which implies that the α values of the interacting atomic orbitals on the two centers are different (α_1, α_2). The two are connected by an interaction integral β. The secular determinant without overlap is a very simple one to write

$$\begin{vmatrix} \alpha_1 - E & \beta \\ \beta & \alpha_2 - E \end{vmatrix} = 0 \tag{7.1}$$

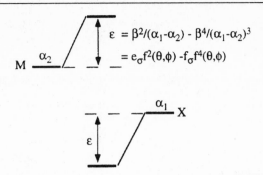

Figure 7.1
The interaction between a pair of orbitals of different energy or electronegativity located on the metal and ligand atoms of an MX_n system evaluated using second-order perturbation theory

It has two roots, $E \sim [(\alpha_1 + \alpha_2) \pm ((\alpha_1 - \alpha_2)^2 + 4\beta^2)^{1/2}]/2$. We can expand under the square root leading to $E \sim \alpha_1 + \beta^2/(\alpha_1 - \alpha_2) - \beta^4/(\alpha_1 - \alpha_2)^3$ and $\alpha_2 - \beta^2/(\alpha_1 - \alpha_2) + \beta^4/(\alpha_1 - \alpha_2)^3$. We know that β, α_1, $\alpha_2 < 0$ and, setting $\alpha_1 < \alpha_2$, the MO diagram of Figure 7.1 results. The stabilization energy of the bonding orbital becomes $\varepsilon = \beta^2/(\alpha_1 - \alpha_2) - \beta^4/(\alpha_1 - \alpha_2)^3$. This is a result that you would get too if you used perturbation theory to evaluate the energy of the interaction. In fact we will use a feature of the perturbation approach below. The term in β^2 is the second-order term and that containing β^4 the fourth-order one.

As in general perturbation theory treatments, presumably the second-order term is larger than the fourth-order one.

Yes. Now the magnitude of the interaction integral β is proportional to the overlap integral between the pair of interacting orbitals and is determined by two factors. The first is the internuclear separation and the second the relative angular disposition of the two orbitals. We will work at constant distance and study the behavior of the angular contribution. This is shown in Figure 7.2 for a hydrogen 1s orbital, or a ligand σ orbital in general, with the metal dz^2 orbital. The dependence of the overlap on geometry is given here by $S = S_0(r)(\frac{1}{2})[3\cos^2(\theta) - 1]$ where r is the interatomic separation. In general $S = S_0(r) f(\theta, \phi)$ where the functions $f(\theta, \phi)$, set by the form of the spherical harmonics that describe[4] the angular behavior of the atomic orbitals, are determined by the polar angles (θ, ϕ) which describe the ligand location for a given axis choice. Using this relationship we can write $\beta = \beta_0 f(\theta, \phi)$ where β_0 contains all of the distance-dependent information. The stabilization energy may now be written $\varepsilon = \beta_0^2 f^2(\theta, \phi)/(\alpha_1 - \alpha_2) - \beta_0^4 f^4(\theta, \phi)/(\alpha_1 - \alpha_2)^3$, or with the introduction of the parameters $e_\sigma = \beta_0^2/(\alpha_1 - \alpha_2)$ and $f_\sigma = \beta_0^4/(\alpha_1 - \alpha_2)^3$, as $\varepsilon = e_\sigma(3\cos^2(\theta) - 1)^2 - f_\sigma(3\cos^2(\theta) - 1)^4$ for the case under consideration. The parameters e_σ, f_σ are thus both determined by the nature of the metal and ligand and the metal–ligand distance. The energy separations of the molecular orbital diagram may then be labeled as shown in Figure 7.1. Remember the e_σ term is the larger of the two and for many purposes orbital diagrams using it alone are good enough.

But one must need a large number of $f(\theta, \phi)$ parameters to describe all the different types of interactions possible.

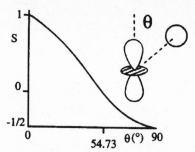

Figure 7.2
The angular behavior of the overlap of a hydrogen 1s orbital, or a ligand σ orbital in general, with the metal dz^2 orbital. (The dependence of the overlap on geometry for this case is $S = S_0(r)(1/2)[3\cos^2(\theta) - 1]$ where r is the interatomic separation)

Although we need to know in general all of the $f(\theta, \phi)$ parameters, for most transition metal complex geometries we actually need only to remember a few. These are shown in Structure **7.3** and are appropriate for the octahedral geometry and all those structures

related to it by removal of ligands; square planar, square pyramidal geometries, etc. We can use this picture to generate the σ molecular orbital diagram for an octahedral MH_6 complex. It's advantageous to use the perturbation theory version of our discussion and focus on the second-order terms only as we just noted. For such a case the approach is remarkably simple. It's just the sum of all of the individual interactions of Structure **7.4**, i.e.,

$$\varepsilon = \sum_i e_\sigma f_i^2(\theta, \phi) \tag{7.2}$$

Thus the energy of interaction of the six ligand σ orbitals ($i = 1$–6) with the z^2 orbital is $(1^2 + 1^2 + (-\frac{1}{2})^2 + (-\frac{1}{2})^2 + (-\frac{1}{2})^2 + (-\frac{1}{2})^2)e_\sigma = 3e_\sigma$. The energy of interaction of the six ligand σ orbitals with the $x^2 - y^2$ orbital is $((-\sqrt{3}/2)^2 + (-\sqrt{3}/2)^2 + (\sqrt{3}/2)^2 + (\sqrt{3}/2)^2)e_\sigma = 3e_\sigma$. Note that the two values are the same, implying that the

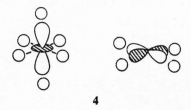

4

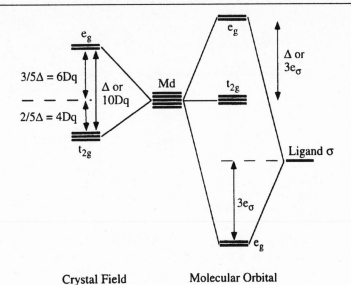

Crystal Field   Molecular Orbital

Figure 7.3
At left the CFT picture of metal-ligand interaction showing splitting of the five atomic d levels. At right the d-orbital, ligand σ orbital only, molecular orbital picture

two orbitals are degenerate, as they must be by symmetry. Of course you can show[4] that you get the same results by writing a normalized symmetry adapted linear combination of orbitals for the e_g ligand sets, and evaluating a single '$f(\theta, \phi)$' for this wavefunction. The xy, xz and yz orbitals have no interaction with the ligand σ orbital orbitals by symmetry and thus remain non-bonding. This d-orbital, ligand σ orbital only, molecular orbital diagram is shown at the right-hand side of Figure 7.3. At the left is shown the corresponding picture from the CFT with its energy label Dq. Note that we use Δ to describe the e_g/t_{2g} splitting in both models.

> It's clear that the AOM is a much more powerful method than the Crystal Field Model for generating molecular orbital diagrams for transition metal complexes since it is easy to evaluate the values of $f(\theta, \phi)$ for any geometry. In many systems of interest there are both σ and π interactions to consider. The CFT is an ionic model and so these labels do not mean anything. How does the AOM accommodate these?

Unlike the CFT, the AOM accommodates σ, π (and δ) interactions quite naturally[3]. The use of the parameters, e_σ and e_π (and e_δ) is a natural extension of our discussion. Structure **7.5** shows the overlap between ligand π and metal $d\pi$ levels. Figure 7.4 shows diagrams that result for octahedral ML_6 and MX_6 complexes containing π acceptors and donors respectively. (A notation, not always adhered to, uses X, L for these types of ligands

5

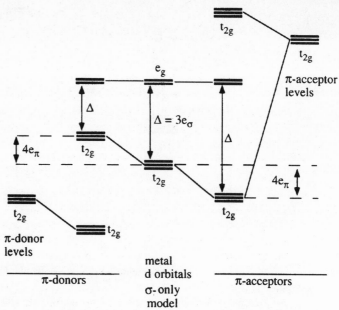

Figure 7.4
Orbital diagrams for octahedral ML_6 ($L = \pi$ acceptor) and MX_6 ($X = \pi$ donor) complexes

respectively.) For this geometry (and all those derived from it by ligand removal) all $f(\theta, \phi) = 1$. The picture shows quite clearly how π acceptors lead to larger, and π donors to smaller, values of Δ. The variation in Δ with the nature of the ligand is described experimentally by the Spectrochemical Series obtained by the study of the electronic spectra of complexes. This is a result of vital importance in understanding the chemistry of these compounds. The CFT is unable to generate such a result.

The chemical control of Δ by the nature of the ligands is indeed an interesting one. Does this control other properties of the compound?

Perhaps the most striking result is the determination of the spin-state of the molecule or solid. (In the latter though, things get more complex because of cooperative effects between adjacent metal atoms.) Structure **7.6** shows high-spin (hs) and low-spin (ls) arrangements for an octahedral d^5 system. Obviously in terms of the one-electron part of the energy, the high-spin arrangement is energetically disfavored (by 2Δ) since two

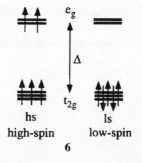

6

electrons reside in high energy orbitals. On the other hand the low-spin form is disfavored since now the electrons are paired up in the same orbital. There will be both Coulomb and Exchange penalties (involving $1/r_{ij}$ terms where r_{ij} is an electron–electron separation) for doing this (see reference 4). Let's call this energy penalty $2P$ (P = pairing energy). Now if $\Delta/P > 1$ the low-spin form will be found and if $\Delta/P < 1$, the high-spin variant will be most stable. For first row transition metals frequently $\Delta/P < 1$ and the high-spin system is found. Often, the two terms are comparable giving rise to interesting temperature- or pressure-induced changes in the nature of the electronic state for these molecules. The second and third row metals are larger (smaller pairing energy since the electrons are further apart on average) and have larger interactions with the ligands (larger Δ) so that invariably $\Delta/P > 1$ and low-spin compounds predominate (see Chapter 4). Another important rôle for π-acceptor ligands will become apparent at the end of this chapter.

There is a sum rule for the second-order angular overlap energies that is useful in assembling molecular orbital diagrams. If there are n ligand λ orbitals ($\lambda = \sigma, \pi$), the sum of the interaction energies over all central atom d orbitals is given by $\Sigma_i \varepsilon_i = n e_\lambda$. As an example, sum the values of ε for each of the e_g orbitals of the octahedral MH_6 complex of Figure 7.3. $\Sigma_i \varepsilon_i = 3e_\sigma + 3e_\sigma = 6e_\sigma$ and six is the total number of σ ligand orbitals. Similarly for the π manifold of Figure 7.4, $\Sigma_i \varepsilon_i = 4e_\pi + 4e_\pi + 4e_\pi = 12e_\pi$ and twelve is the total number of ligand π orbitals. The sum rule is particularly useful in generating the level diagram, and thus the value of Δ, for the tetrahedral molecule. We know from group theory that the d orbitals split as in Structure **7.7**, and that the t_2 set only are involved in σ

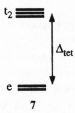

7

interactions and so lie higher in energy. Thus the value of ε for the σ orbitals is just $4e_\sigma/3$ since there are four σ ligands and three orbitals. On this σ-only model therefore, $\Delta_{tet} = 4e_\sigma/3$. By comparison with the results for the octahedron ($\Delta_{oct} = 3e_\sigma$), you can see that $\Delta_{tet} = \frac{4}{9}\Delta_{oct}$. The result is more complex if π interactions are included; the factor of $\frac{4}{9}$ disappears. But remember that these interactions are less potent than σ ones.

This is a much more rapid (and more elegant) way to derive this ratio than using the CFT. Is there a molecular orbital analog of the CFSE?

As you noted one of the immediately accessible results of the crystal field theory is the correlation between the heats of formation of transition metal systems and the crystal field stabilization energy (CFSE). There is indeed an analogous way of viewing the stabilization energy of complex formation as a function of d count but using the AOM. It introduces the idea of the Molecular Orbital Stabilization Energy (MOSE). On formation of the complex, from Figure 7.3 we see that four electrons are stabilized in the deepest-lying orbital leading to a MOSE of $4 \times 3e_\sigma = 12e_\sigma$. On filling the t_{2g} orbitals this does not change since on the σ-only model these levels are non-bonding. However, filling the e_g levels is energetically penalizing and the MOSE is reduced. We can generalize this by writing

MOSE $= 12e_\sigma - n(e_g)e_\sigma$ where $n(e_g)$ is the number of electrons in the e_g orbital set. Figure 7.5 shows how the MOSE on this d-orbital only model varies with d count. Also shown is the corresponding variation in CFSE. Both plots correspond to high-spin situations. Addition of a judiciously chosen sloping background in each case in Figure 7.6 leads to identical plots. Again this is a purely symmetry-based result for both models. Similar results are obtained from the AOM if π effects are included since in general $e_\sigma > e_\pi$. Figure 7.7 shows a plot of some experimental data. Notice that the variation across the series on both models is small compared to the total energy.

Where does the sloping background come from?

The origin of the energetics of compound formation is usually difficult to pinpoint (see Chapter 5). However, using the electrostatic model behind the CFT, the positive slope may come from the gradual decrease in metal–ligand distances on moving from left to right across the Periodic Table. From the molecular orbital approach, the small contribution to ΔH from these nd orbital effects that shows up in Figure 7.7 emphasizes the importance of the metal $(n+1)$s, p orbitals in determining these results. The slope then represents the increasing importance of these orbitals on moving across the Periodic Table. By the time we reach zinc the d orbitals are effectively part of the core.

And you can use the same type of approach to study labile and inert transition-metal complexes?

Yes, one nice application of these d-orbital-only ideas is in the study of labile and inert transition metal complexes as you noted. Complexes of Cr(III) react slowly in water and are regarded as 'inert', but complexes of Cr(II) are labile, rapidly losing the ligands from the coordination sphere. We can easily compute the d count variation in the activation energy, which we will identify as being due to the change in CFSE or MOSE on going from octahedron to square pyramid. (This may not actually be the mechanism here but certainly breaking one metal–ligand bond will play a part.) The octahedral diagram was generated in Figure 7.3. The corresponding picture for the square pyramid using equation (7.2) is shown in Figure 7.8. This is a really easy result to derive using the AOM.

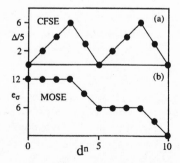

Figure 7.5
(a) The variation in CFSE with d count for an octahedral complex; (b) the analogous variation in MOSE (high-spin configurations were chosen here)

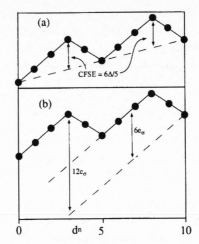

Figure 7.6
(a), (b) Addition of a sloping background to the plots of Figure 7.5

So the energy of interaction of the five ligand σ orbitals ($i = 1$–5) with the z^2 orbital is $(1^2 + (-1/2)^2 + (-1/2)^2 + (-1/2)^2 + (s - 1/2)^2)e_\sigma = 2e_\sigma$. The energy of interaction of the five ligand σ orbitals with the $x^2 - y^2$ orbital is $((-\sqrt{3}/2)^2 + (-\sqrt{3}/2)^2 + (\sqrt{3}/2)^2 + (\sqrt{3}/2)^2)e_\sigma = 3e_\sigma$, i.e., the same result as for the octahedron.

It is much more difficult to derive this result using the CFT. First the geometry is non-cubic and requires the use of a second parameter Cp in addition to Dq (although most people don't realize this). Second we don't know the ratio between the two. However, once a choice of $\rho = Cp/Dq$ has been made, the energy differences $M(H_2O)_6 \rightarrow M(H_2O)_5 + H_2O$, ΔMOSE and ΔCFSE are easy to calculate from the AOM and CFT diagrams of the octahedron and square pyramid. The results with a sloping

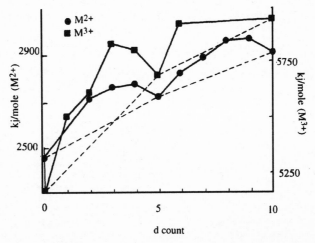

Figure 7.7
The experimentally determined heats of hydration for the first row divalent (circles) and trivalent (squares) transition metal ions. Notice the value of the origin of the abscissa

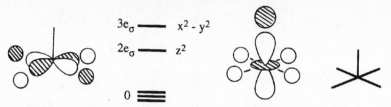

Figure 7.8
The AOM orbital diagram for the square pyramid

background added, correlate[4] nicely with the observed log (rate constants) as shown in Figure 7.9.

So, much of the success of these models is based on symmetry. The result concerning the lability of transition metal complexes is set by the form of the MX_6 and MX_5 energy level diagrams. In this sense, for problems of this type it doesn't really matter which model you use. To understand the origin of the variations in Δ for different ligands in terms of σ and π bonding, the greater flexibility of the AOM is vital. Although it is a very simple model it does capture much of the essence of transition metal–ligand interaction. Numerical details of bond lengths, bond energies, and the importance of s/p/d mixing of course have to be tackled using more sophisticated models.

And so what's behind the Eighteen Electron Rule? It works well for compounds such as $Cr(CO)_6$ but is certainly not applicable to $Ni(H_2O)_6^{2+}$.

Its origin is relatively simple. It's sometimes called the Effective Atomic Number (EAN) rule. In this case the total number of electrons (formally) required for stability is that corresponding to the next Group 18 ('Inert') gas, in this case $2(s) + 6(p) + 10(d) = 18$.

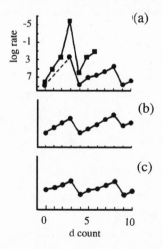

Figure 7.9
(a) The observed log (rate constant) for water loss from divalent (circles) and trivalent (squares) ions of the first-row transition metal ions; (b), (c) the computed d count variation in the activation energy for ligand loss from an octahedral complex, assumed to be equal to the change in: (b) MOSE; or (c) CFSE ($\rho = 1$) on going from octahedron to square pyramid, with the addition of a sloping background

Let's first look at the molecular orbital diagram for an octahedral MH_6 complex using d, s and p orbitals on the metal. This is shown in Figure 7.10 and is constructed simply by pairing ligand orbital combinations of a given symmetry species with their central atom counterparts. Notice that there are six bonding orbitals (of $a_{1g} + e_g + t_{1u}$ symmetry) and three non-bonding orbitals (of t_{2g} symmetry). Using the rule that stable systems arise when all the bonding and non-bonding levels are occupied, a total of eighteen electrons results. If the ligands are π-acceptors in addition, then the t_{2g} set is stabilized and all of the occupied levels are bonding ones. Notice that our criterion of stability also implies that there is a good HOMO–LUMO gap in the molecule. In many systems a large gap of this type leads to geometric and kinetic stability. If the ligands are weak field ones and especially if they are π-donors in addition then the result will be a small e_g–t_{2g} gap and the e_g, σ-antibonding orbitals can be occupied too, with little energetic penalty, thus invalidating the rule. Bond lengths will increase a bit but the molecule will not fall apart. Your example of $Ni(H_2O)_6^{2+}$ falls into this category. Thus for the octahedral complex, eighteen electrons only lead to special stability when the metal is coordinated by strong-field ligands.

So, since antibonding orbitals are found at high energy by definition in these systems and thus lead to large gaps between bonding and antibonding levels, stable compounds form when no strongly antibonding orbitals are occupied.

Yes. The result may be generalized. In many other geometries we find that if there are n ligand σ orbitals then the molecular orbital diagram will contain n bonding orbitals and n antibonding orbitals to high energy. With five d, three p and one s orbital on the central

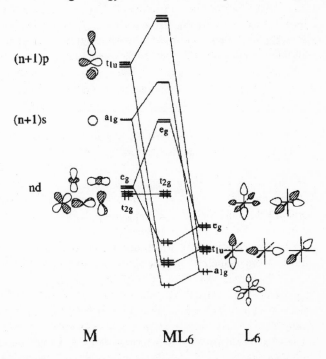

Figure 7.10
A molecular orbital diagram for an octahedral MH_6 complex using the valence d, s and p orbitals on the metal

metal atom there will then be $9 - n$ non-bonding orbitals. Filling all bonding and non-bonding orbitals leads again to a total of $2(n + 9 - n) = 18$ electrons associated with the metal center. We should realize that although we talk of 18 electrons 'associated with the metal', some of these lie in orbitals that are largely ligand-located (see Figure 7.10). This counting scheme is just a way of keeping track of the electrons.

How do you count the electrons?

The easiest way is to fill all of these deep-lying ligand orbitals with electrons. Thus Cl^- contributes two electrons from a metal-directed s/p hybrid orbital, as does CO or ethylene or CN^-. We usually assume that η^n rings of cyclic polyenes with n atoms contribute the number of electrons appropriate for their Hückel rule stability as shown in Structure **13.1**. Thus η^5-C_5H_5 (Cp) we count as Cp^- which contributes six electrons. So ferrocene, Cp_2Fe, has an eighteen electron count if we regard it as $Cp_2^- Fe^{2+}$ ((2×6) from Cp^- and 6 from Fe^{2+}). This method is quite arbitrary; other approaches are just as good. The molecule still has eighteen electrons if we build it from neutral fragments as Cp_2Fe ((2×5) from $Cp + 8$ from Fe).

A similar scheme is presumably behind the eight-electron rule as well.

Yes, both eight- and eighteen-electron rules have the same electronic origin. The Effective Atomic Number Rule describes both. Whereas s, p and d orbitals are formally filled in eighteen electron systems, s and p only are filled in eight electron (octet) systems. So methane has four bonding pairs of electrons, ammonia has three bonding pairs plus one lone pair (just as $Cr(CO)_6$ has six σ bonding pairs plus three σ non-bonding pairs) arising from the four (one s and three p) orbitals on the central atom. There are no truly twenty-electron compounds, i.e., molecules where high energy orbitals are occupied. In $Cr(CO)_6$, for example, if we promote an electron into the e_g orbital photochemically, the molecule loses CO to give the sixteen-electron compound $Cr(CO)_5$, which will then combine with that CO or some other species nearby to regenerate an eighteen electron molecule.

But what about $W(CO)(C_2H_2)_3$? Counting electrons here leads to a total of twenty. I also know of stable compounds with less than eighteen electrons.

This is an interesting case. The electron count is right but there is one ligand combination (Structure **7.8**) which by symmetry (a_2) cannot interact with a central atom orbital[3]. We looked at similar main group 'hypervalent' molecules in Chapter 6. Although ferrocene is a very stable eighteen-electron compound, the molecule nickelocene with twenty electrons is also known. In this geometry there is a low-lying orbital that may be filled. More numerous are compounds with fewer than eighteen electrons. These are systems where not all of the non-bonding orbitals are filled. Seventeen-electron $V(CO)_6$ is an example. The usual explanation for its stability is that generation of an eighteen-electron count leads to a steric congestion. $V(CO)_6$ would have to form the dimer, $V_2(CO)_{12}$, to satisfy the eighteen-electron rule at each metal center and CO...CO repulsions could be large between the fragments thus discouraging its formation. Such a problem does not exist with the seventeen-electron fragments $Mn(CO)_5$ or $Co(CO)_4$. They dimerize to give $Mn_2(CO)_{10}$ and $Co_2(CO)_8$ respectively. Often these molecules with smaller electron counts contain agostic hydrogen atoms. These are hydrogen atoms,

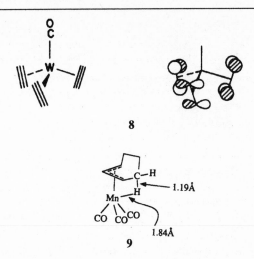

8

9

attached to the organic part of an organometallic molecule that get close to the metal and formally increase the electron count by two. Structure **7.9** shows an example where the agostic M–H linkage increases the electron count at Mn from sixteen to eighteen. These extra two electrons formally lie in a C–H bonding orbital of the organic unit before coordination. The result is an increase in the C–H distance (to 1.19 Å in this case) on formation of the agostic linkage.

> You haven't mentioned square planar molecules such as $Pt(CN)_4^{2-}$ that have a sixteen-electron count.

This corresponds to the opposite situation to the one in $W(CO)(C_2H_2)_3$. In the square planar geometry one of the p orbitals (that perpendicular to the plane, Structure **7.10**) is of the wrong symmetry to interact with any of the ligand σ orbitals. Also, it lies too high in energy to be filled. Thus the number of available metal orbitals here is $8 - n$ and the stable electron count is sixteen. A similar situation is found for planar three-coordinate main group molecules. BX_3 molecules (X = halide) with six electrons are perfectly stable. Here analogously, the p orbital perpendicular to the plane (Structure **7.11**) is of the wrong symmetry to interact with the ligand σ set. There are some nice comparisons between the two types of compound. BX_3 halides readily add halide ion to give a tetrahedral eight-electron compound (equation (7.3)) just as square-planar sixteen-electron compounds add a fifth or sixth ligand to give an eighteen-electron compound. Equation (7.4) shows an oxidative addition reaction.

$$BX_3 + X^- \rightarrow BX_4^- \tag{7.3}$$

$$trans - Ir^I Cl(CO)(PPh_3)_2 + HCl \rightarrow Ir^{III} HCl_2(CO)(PPh_3)_2 \tag{7.4}$$

10

11

References

1. Dunn, T. M., McClure, D. S. and Pearson, R. G., *Some Aspects of Crystal Field Theory*, Harper and Row (1965).
2. Gerloch, M. and Slade, R. C., *Ligand Field Parameters*, Cambridge University Press (1973).
3. Albright, T. A., Burdett, J. K. and Whangbo, M.-H., *Orbital Interactions in Chemistry*, John Wiley & Sons (1985).
4. Burdett, J. K., *Molecular Shapes, Theoretical Models of Inorganic Stereochemistry*, John Wiley & Sons (1980).

General References and Further Reading

Ballhausen, C. J., *Introduction to Ligand Field Theory*, McGraw-Hill (1964).

Ballhausen, C. J. and Gray, H. B., *Molecular Electronic Structure*, Benjamin-Cummings (1980).

Bersuker, I. B., *Electronic Structure and Properties of Transition Metal Compounds*, John Wiley & Sons (1996).

Bethe, H., *Ann. Phys.*, **3**, 133 (1928).

Cotton, F. A., *Chemical Applications of Group Theory*, 3rd Edition, John Wiley & Sons (1990).

DeKock, R. and Gray, H., *Chemical Structure and Bonding*, Benjamin/Cummings (1980).

Gerloch, M., *Coord. Chem. Rev.*, **99**, 117 (1990).

Jorgensen, C. K., *Modern Aspects of Ligand Field Theory*, North-Holland (1971).

Jorgensen, C. K., Pappalardo, R. and Schmidtke, H. H., *J. Chem. Phys.*, **39**, 1422 (1963).

Orgel, L. E., *Introduction to Transition Metal Chemistry*, Second Edition, Methuen (1966).

Schäffer, C. E., *Structure and Bonding*, **14**, 69 (1973).

Smith, D. W., *Structure and Bonding*, **12**, 49 (1972).

8 Are Main Group Molecules and Transition Metal Complexes not so Different after all?

Isn't it striking that the structure of $Os_2(CO)_6C_2R_2$ (Structure **8.1**) contains the same square geometry for the heavy atoms as cyclobutane (Structure **8.2**). It is as if in some way the fragment $Os(CO)_4$ is similar electronically to CR_2. Similarly the molecule $Co_4(CO)_{12}$ (Structure **8.3**) has the same basic shape as the tetrahedrane derivative C_4R_4 (Structure **8.4**). It is as if the CR unit is similar electronically in some way to that of $Co(CO)_3$. How general are these results?

It's interesting that you used the terms molecule and complex to describe species from the two parts of the Periodic Table. It's associated with the different historical development of the two areas. But, yes indeed they are similar. Roald Hoffmann has called this relationship the isolobal analogy[1], and much of the present interest in the scheme is due to him (but see reference 2 as well). Two fragments (such as CR and $Ir(CO)_3$) are isolobal to each other if the basic shape and orbital occupation of their frontier orbitals are similar. The analogy provides a nice connection between the orbitals of inorganic or organometallic fragments and those in the organic and main group areas, as you pointed out. The idea has its origins in Jack Halpern's insights concerning similarities between organic and transition metal intermediates and the structural stability patterns of the cage and cluster molecules typified by the boranes and transitional metal clusters that were unearthed by Kenneth Wade[3] and Michael Mingos[4] that we discuss in Chapter 11. Our discussion here comes from reference 5.

Figure 8.1 shows a simple way to see how the isolobal relationship links various main group and organometallic fragments. (An isolobal relationship is identified by a double-

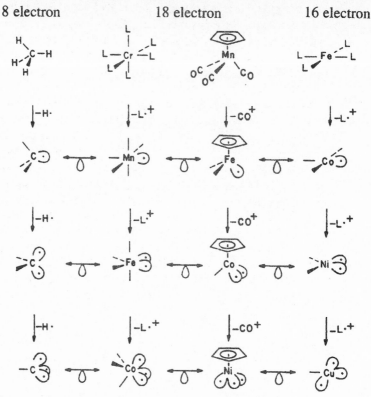

Figure 8.1

The origin of the isolobal relationship between main group and organometallic fragments (adapted from reference 5)

headed arrow with a tear-drop, meant to represent an orbital.) It is based on a valence bond type of idea, that an eight-electron tetrahedral molecule such as methane can be envisaged as being composed of four hybrid orbitals pointing toward the corners of a tetrahedron, and an eighteen-electron octahedral molecule such as $Cr(CO)_6$ envisaged as containing six hybrids pointing toward the corners of an octahedron (Structure **8.5**). We know what happens on breaking one C—H bond in methane in a homolytic sense. The methyl radical is generated with a frontier orbital, containing one electron, pointing in the direction of the missing hydrogen atom. Loss of two hydrogen atoms leads to methylene with two hybrid orbitals each containing an electron pointing toward the two missing hydrogens. In an exactly analogous way, loss of one more hydrogen atom leads to a methine fragment with three frontier orbitals of the same type. Although we have used methane in this example we could have started with ammonia. Loss of one hydrogen atom then leads to an NR_2 unit

5

with one hybrid orbital containing two electrons and one containing just one. Similarly we could have used the ammonium ion, NH_4^+, instead of methane as our starting molecule and derived the orbitals of NH_3^+, NH_2^+ and NH^+. How these fragments can be bound together to form a huge variety of stable molecules with eight electrons at each center is well known. One just forms two-center two-electron bonds between pairs of atoms using the unpaired electron in each hybrid.

Exactly the same process may be used for $Cr(CO)_6$ or any other d^6 transition metal based molecule with six two-electron ligands leading to an eighteen-electron complex. Breaking one $Cr-CO$, or $Cr-L$ linkage in general (where L is a two-electron ligand), in a homolytic fashion, just as we did for methane, leaves one electron on L and one with Cr. So, a CrL_5^- fragment is produced along with a positively charged L^+ ligand. To avoid having to remember how many charges we have at the metal let's make things a bit easier for ourselves by using the fact that neutral Mn is isoelectronic with Cr^-. Notice that the result is then a square pyramidal d^7, MnL_5 unit. Notice too that it has one hybrid orbital pointing toward the missing ligand, and that this orbital contains one electron. This is just the situation found for CH_3. The orbital composition of this hybrid is a mixture of metal 3d with some metal 4s and 4p character, just as the corresponding orbital in CH_3 is a mixture of carbon 2s and 2p character. Importantly it looks just like the frontier orbital of the methyl radical. As a result we say that the fragments MnL_5 and CH_3, or CR_3 in general, are isolobal.

We noted earlier that we didn't need to stick just to methane as the starting compound; we could just have as easily started with the ammonium ion. Is the same is true of the transition metal-containing fragments?

Yes, any square pyramidal d^7 unit will do, derived from an eighteen-electron parent. Thus $Co(CN)_5^{3-}$ is also isolobal with CH_3. Having one unpaired electron in an outward-pointing hybrid implies that both $Co(CN)_5^{3-}$ and $Mn(CO)_5$ will behave structurally and chemically like CH_3.

Let's continue this ligand removal process. Taking off another ligand (hydrogen atom) from CH_3 leads of course to CH_2. Removing a ligand from MnL_5 and moving one element to the right in the Periodic Table to keep care of the charges as we did earlier, leads to the FeL_4 fragment with a structure described as an octahedral cis-divacant, or sawhorse geometry. (We will see in a minute why we didn't choose the square geometry.) Clearly it is isolobal with CH_2 or NR_2^+. Since Fe is clearly isoelectronic with Os, an element below it in the Periodic Table, $Os(CO)_4$ is isolobal with CR_2 too. So here briefly is the chemistry behind the observation in Structures **8.1** and **8.2** that the two molecules are structurally quite alike. The spatial behavior of the frontier orbitals, and the numbers of electrons in them, are similar. Removal of another ligand leads to the pyramidal CoL_3 fragment that naturally is isolobal to CH or CR. Once again we can see how the structural similarity of the two molecules of Structures **8.3** and **8.4** is no accident. At its simplest level, if you can understand how the tetrahedrane molecule holds together by linking together in pairs the unpaired electrons in the three hybrids of each CH fragment, then you can understand how the $Co_4(CO)_{12}$ molecule holds together by using the orbitals and electrons of the analogous $Co(CO)_3$ fragment.

Can we choose another geometry and composition for our starting eighteen-electron rule compound?

Although we started with the octahedral CrL_6 molecule as our quintessential eighteen-electron transition metal complex, we could certainly have chosen other species. The molecule cymantrene $(\eta^5\text{-}C_5H_5)Mn(CO)_3) = CpMn(CO)_3$, which contains a coordinated cyclopentadienyl molecule, is another interesting place to start. Thus $CpFe(CO)_2$ is isolobal with CR_3, $CpCo(CO)$ isolobal with CR_2, and $CpNi$ isolobal with CR. You can generate in a similar way the fragments that may be derived in an analogous way from $(\eta^6\text{-}C_6H_6)Cr(CO)_3$.

The isolobal possibilities must be enormous.

Yes, Cr, for example, is just as good as the isoelectronic species Mo, W, V^-, Nb^-, Ta^-, Mn^+, Tc^+, Re^+ etc. Similarly CR_3 is just as good as $Ga(CH_3)_3^-$ or $N(C_2H_5)_3^+$. CO is a two-electron ligand as are the phosphines PR_3. Thus, CH_2 is isolobal to C_{2v} $Fe(CO)_4$, $Ru(PR_3)_4$, or $Ir(CO)_4^+$, and $Sn(CH_3)_3$ is isolobal to $Mn(CO)_5$, $Os(PR_3)_5^+$, or $Mo(CO)_5^-$ as well as to $CpMn(CO)_2^-$. You can write down many more with the aid of the Periodic Table. Of course there may be some changes in the geometries of the fragments when compared to the structures found for these units in the parent.

You avoided the square planar complex. Is there something special about it?

No, This is just a reflection of the fact that there are stable electron counts other than the eighteen-electron one. Thus the square planar complex in Figure 8.1, is stable for 16 electrons. The example usually used here is that of $Pt(CN)_4^{2-}$ or FeL_4. We can proceed in exactly the same way as before to show that the T-shaped CoL_3 unit is isolobal with CR_3 and the bent NiL_2 unit isolobal with CR_2.

Each time you have used a starting compound that obeys one of the 'closed shell' rules, sixteen or eighteen electrons. Do these ideas hold when this rule is not obeyed?

Indeed we need to use compounds that obey these rules in order to be able to make comparison with the compounds arising from octet rule chemistry. If you recall our discussion concerning the background to the EAN rules of Chapter 7, the types of ligands in the transition metal-containing systems that are used, must be those which lead to large values of Δ.

What about chemistry?

There are quite extensive comparisons. Not only, as we have seen already, do these isolobal fragments form compounds of similar type, they very often have similar properties. Thus the species $Mn(CO)_5$ has one unpaired electron and behaves like CR_3 with one unpaired electron. Both readily dimerize and abstract H and Cl from other molecules. Since CR_3 is isolobal with $Mn(CO)_5$, replacement of one or both CR_3 groups in ethane, C_2R_6, leads to $CR_3Mn(CO)_5$ and $Mn_2(CO)_{10}$. In each there is a single bond, C—C, C—Mn and Mn—Mn respectively. Molecular orbital diagrams for all three molecules are similar (Figure 8.2). Photochemical excitation of $Mn_2(CO)_{10}$ leads to population of the σ^* orbital, photodissociation and generation of $Mn(CO)_5$ radicals. Similar excitation of ethane leads to methyl radicals. The two unpaired electrons in CH_2 and $Fe(CO)_4$ lead to singlet and triplet electronic states in both cases (Structure **8.6**). The geometries of the two states and the singlet–triplet energy difference have received much attention. The bond angles in the two methylene molecules are quite different and, while the geometry of singlet $Fe(CO)_4$ is not known, the triplet has a distorted octahedral cis-divacant structure. The conclusions for CH_2 come largely from the gas phase, whilst those for $Fe(CO)_4$ come from matrix isolation studies. The triplet is lower in energy for both $Fe(CO)_4$ and CH_2. For CF_2 the singlet is lower in energy.

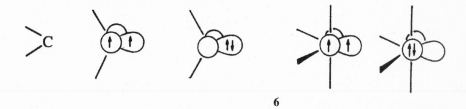

6

Let's look now at some ways other than those of Structures **8.1** and **8.2** to make isolobal analogies with CR_2. Recall that $Fe(CO)_4$ is isolobal with CR_2. Thus ethylene-$Fe(CO)_4$, $(CR_2)_2Fe(CO)_4$ (Structure **8.8**), is equivalent to cyclopropane, Structure **8.7**. In fact the way it is drawn emphasizes the metallacyclopropane viewpoint for this molecule and highlights the octahedral coordination at iron. The other view of ethylene coordination is shown in Structure **8.12**, to be compared with the new view of cyclopropane in Structure **8.11**. These stress extreme descriptions of the chemical bonding that may be envisaged for this species. Very similar structures are found with other organometallic fragments isolobal to $Fe(CO)_4$. So, for example, there are analogous compounds of ethylene found for $Ni(PPh_3)_2$ (Structures **8.9** and **8.13**) and $CpCo(CO)$ (Structures **8.10** and **8.14**).

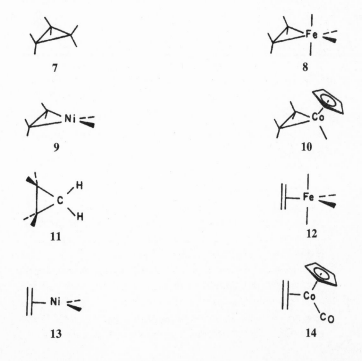

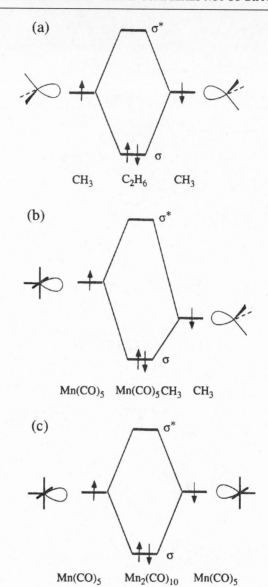

Figure 8.2
A part of the molecular orbital diagrams for C_2R_6, $CR_3Mn(CO)_5$ and $Mn_2(CO)_{10}$

Your approach has used a localized model and is very effective, but what about a delocalized one?

Although we have used the localized model in Figure 8.1, the delocalized picture does give added insight here. Thus Figure 8.3 shows the orbitals of CH_2 and $Fe(CO)_4$ and their interaction with ethylene to give respectively cyclopropane and (ethylene)$Fe(CO)_4$. Notice that although the ordering of the frontier orbitals is different in the two fragments, the final pictures are very similar. Both localized and delocalized pictures give analogous

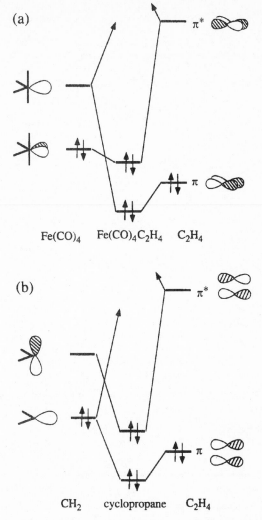

Figure 8.3
Orbital interaction diagrams for: (a) Fe(CO)$_4$ and ethylene; (b) CH$_2$ and ethylene

predictions (Structures **8.8** and **8.12**) concerning the orientation of the ethylene molecule in Fe(CO)$_4$(C$_2$H$_4$).

A further set of isolobal examples is shown in Structures **8.15– 8.17**. Noting the isolobal analogy between Os(CO)$_4$, CpRh(CO) and CH$_2$ units, and remembering that tin and carbon lie in the same column of the Periodic Table, leads to the result that the molecules Structures **8.15** and **8.16** are isolobal with spiropentane, Structure **8.17**.

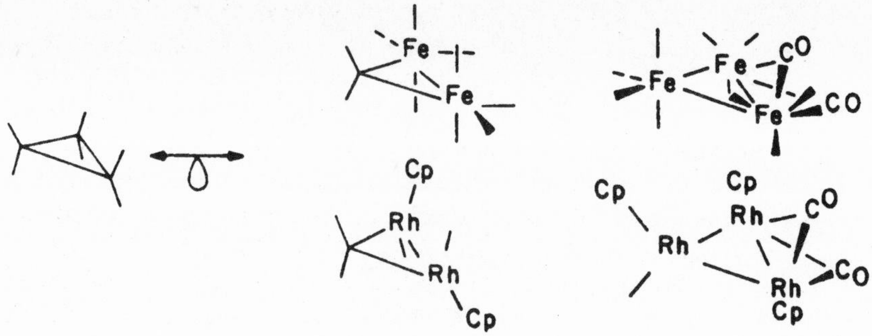

18

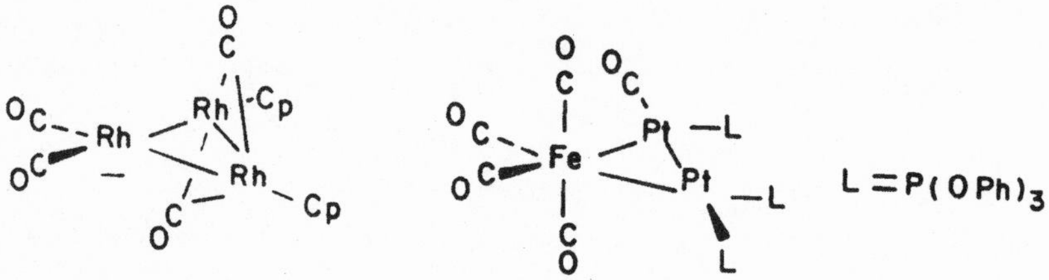

19

If we replace first two and then all three CH_2 groups in cyclopropane then we find the set of compounds of Structure **8.18**. Two interesting 'cyclopropanes' containing ML_2 units derived from a square planar starting structure are shown in Structure **8.19**. All of these are known species. However, there is an interesting twist here. The structure of $Fe_3(CO)_{12}$ is more complex than we might have imagined. It contains two bridging carbonyl groups. In the isoelectronic $Ru_3(CO)_{12}$ and $Os_3(CO)_{12}$ species though, all the carbonyl groups are located in terminal positions. Such bridging CO groups are very common for first-row transition metal systems but are not unknown for the heavier elements. In terms of electron counting it makes no difference. The two electrons of each CO group are shared between two metal atoms. Since the bridging groups in $Fe_3(CO)_{12}$ occur as a pair, the electron count at each center remains the same. The propensity of the first-row metal atoms to indulge in this sort of structure is thought to be due to short metal–metal distances. This allows the

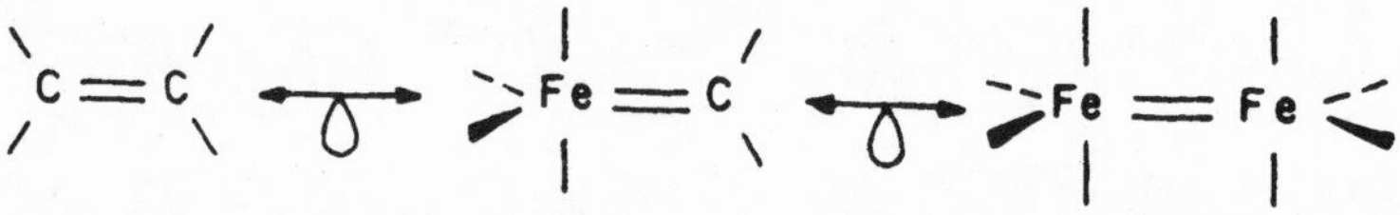

20   **21**

CO group to easily span the metal–metal bond, whereas in the heavier analogs the metal–metal distance is often too long to allow binding of the CO to both metal atoms. In fact your molecule $Co_4(CO)_{12}$ of Structure **8.3** also contains bridging groups. Only as its heavier analog $Ir_4(CO)_{12}$ is the all-terminal isomer found.

Does the analogy always lead to the prediction of stable transition-metal-containing compounds where stable main group ones exist?

No, the analogy does not always lead to the prediction of stable structures. Thus although two CH_2 units lead to the stable ethylene molecule and, although the species Structure **8.20** can be made, the molecule Structure **8.21** has been characterized only in a low-temperature matrix and is unstable at higher temperatures. This brings up a long-standing aspect of chemical bonding. Although we can devise many molecules that should be thermodynamically stable using rules such as these, the question of their kinetic stability is one that is usually difficult to answer. One from octet chemistry is that of tetrahedrane (C_4H_4) and its derivatives (Structure **8.4**), of relevance to some of the molecules in this chapter. There is nothing wrong with it in terms of electron counting since each carbon atom has an octet of electrons. However, the molecule is not known at present, although derivatives such as $C_4{}^tBu_4$ have been synthesized. The hydrogen atoms in the parent are expected to be highly acidic given the strained environment at carbon, and the molecule in general will be open to nucleophilic attack. The ${}^tBu$ groups remove the hydrogen acidity problem and act as a sheath to protect the C_4 core from attack. Thus a kinetically unstable molecule may be stabilized by derivatisation.

But let's continue. We showed above how a CH fragment is isolobal to $Co(CO)_3$. Although you noted the analogy between tetrahedrane and $Co_4(CO)_{12}$, there is a whole set of compounds shown in Structures **8.22–8.26** where $Co(CO)_3$ groups have been replaced by CR units. All are known molecules. These species do lead us to other ways to describe these molecules. The molecule in Structure **8.23** is usually drawn as in Structure **8.27**, namely as a cyclopropenium complex. The compound in Structure **8.24** is usually described as a $Co_2(CO)_6$ unit bridged by an acetylene molecule (note that the C–C linkage lies perpendicular to the Co–Co linkage (Structure **8.28**) as we have just noted. Just as in the case of $Fe_3(CO)_{12}$ the Co compound, Structure **8.26**, actually has bridging CO groups. The heaviest isoelectronic species, $Ir_4(CO)_{12}$ does have the structure shown with all CO groups terminal.

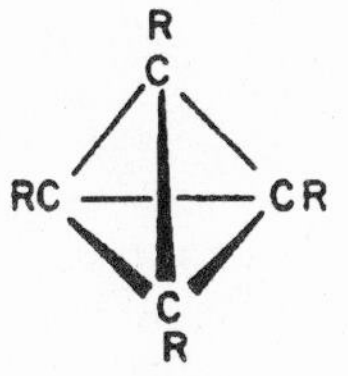

22

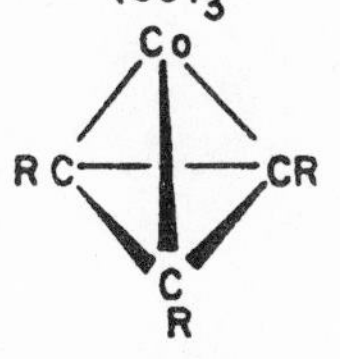

23

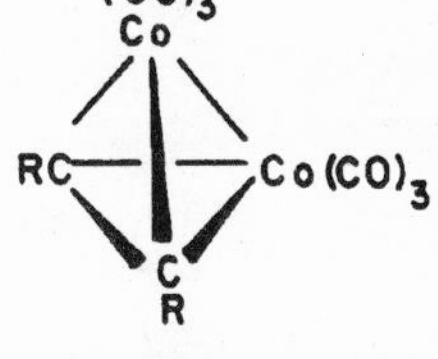

24

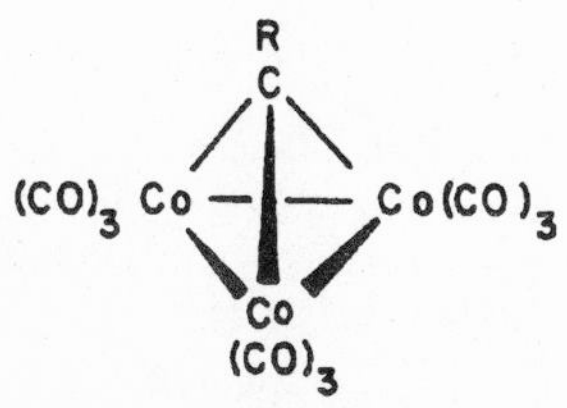

25

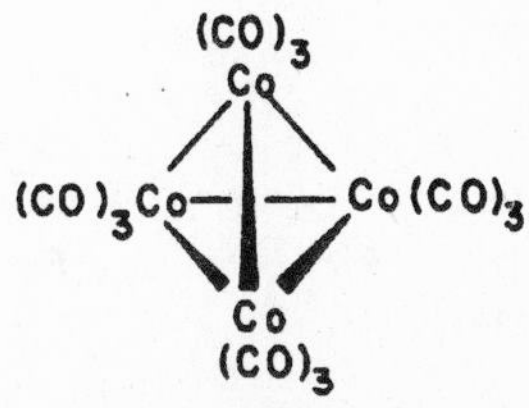

26

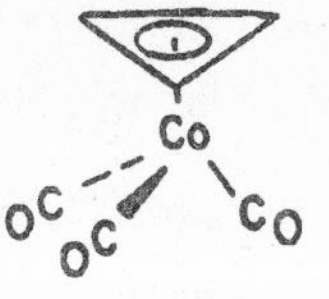

27

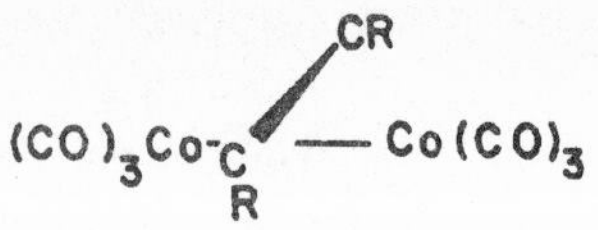

28

Another interesting set of compounds results when the ML_3 fragment is substituted by an isolobal fragment in the metallocenes $(\eta^n C_n H_n)ML_3$. For example an analog of cyclobutadiene-$Fe(CO)_3$, Structure **8.29**, is the $C_5H_5^+$ cation. Since $Mn(CO)_3$ and CH_2 fragments are isolobal, then cymantrene is isolobal with the benzene dication Structure **8.30**. The crystal structure of the $C_6H_6^{2+}$ derivative shown in Structure **8.31** has been determined. Tin, of course is isoelectronic with carbon. Interestingly the apical CR^{2+} unit

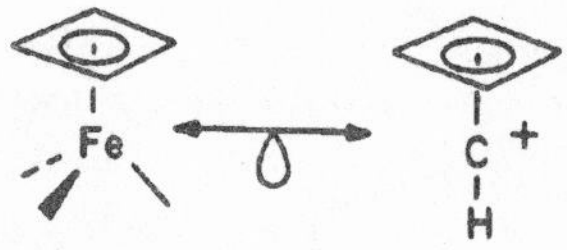

29

30

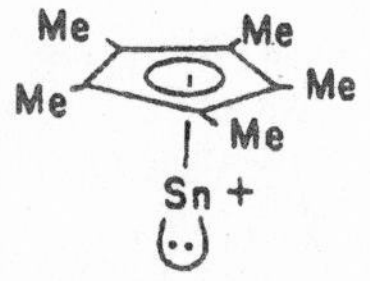

31

in this molecule may be replaced by the isoelectronic $P-Me^+$ unit to give a stable compound with the same structure. Note that Structure **8.30** contains a 'six-valent carbon' and Structure **8.29** a 'five-valent' one.

These are pretty unusual coordination numbers for carbon aren't they?

Actually, coordination numbers larger than four are frequently found in carbocations. In addition, the element adjacent to it, boron, forms many compounds where the coordination number is five. Some very interesting compounds in inorganic chemistry are in fact those which contain cages or clusters of atoms. Structure **8.32** shows examples of such molecules. We shall see how to count electrons for these species in Chapter 11. In these molecules one can readily see how the isolobal replacement of a BH by a CoCp unit leads to an interesting new species.

You have mentioned the eight- and eighteen-rule analogies. What about the 'doublet rule'?

I suppose you mean here analogies with the hydrogen atom where there is only one atom or orbital involved. Structure **8.33** shows a nice equivalence between the molecules H_3^+, $W(CO)_2(PR_3)_3(\eta^2 H_2)$ and CH_5^+ ($CH_3(H_2)$). H^+ is thus isolobal with d^6 square pyramidal ML_5 and CR_3^+ species. Another useful analogy links the frontier orbitals of $H \cdot$ and MPR_3 units ($M =$ Group 11 atom). Structure **8.34** shows the origin of the comparison in a simple

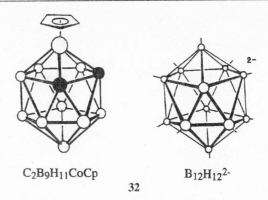

$$C_2B_9H_{11}CoCp \qquad B_{12}H_{12}{}^{2-}$$

32

way. Figure 8.4 shows orbital diagrams for the H_4 tetrahedron (see Chapter 2) and H_6 octahedron with closed shell stability for $H_4{}^{2+}$ and $H_6{}^{4+}$ (two unknown species). The molecule $(Au(PR_3))_4I_2$, containing the $(Au(PR_3))_4{}^{2+}$ ion however, is known. It is isoelectronic with $H_4{}^{2+}$. Of interest is the result of inserting a small atom, O, N, C, into the

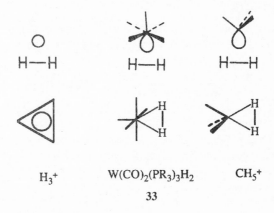

$$H_3{}^+ \qquad W(CO)_2(PR_3)_3H_2 \qquad CH_5{}^+$$

33

cluster. Figure 8.5 shows how the orbital diagram changes for the octahedron. (A similar picture will be found in Figure 11.2.) Now the structure is stable for eight electrons. The ion $(Au(PR_3))_6C^{2+}$ is known with this structure.

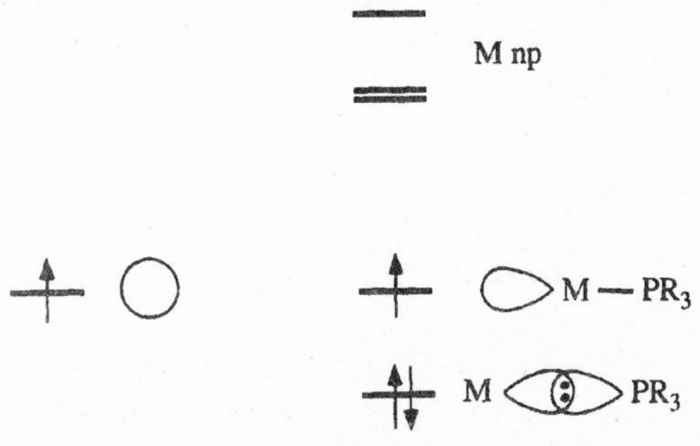

M np

H

MPR$_3$

34

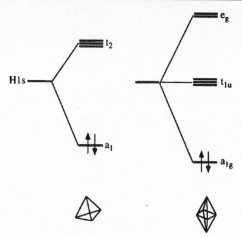

Figure 8.4
Orbital diagrams for the H_4 tetrahedron and H_6 octahedron

It is clear from the derivation of the analogy in general that it only links doublet, octet and eighteen electron compounds as we discussed above. In the transition metal case these are compounds that have large values of Δ (the e_g/t_{2g} splitting in octahedral complexes) and invariably contain organometallic ligands. So at this junction between the transition metal and main group series, for gold chemistry as the d block is sinking into the core, rather different considerations often come into play with these Group 11 compounds. Here not only the metal s, as we have described, but also a variable number of the metal p orbitals, are often important. The result is that the chemistry of many MPR_3 systems is very different[6] from what is predicted using the approach described above.

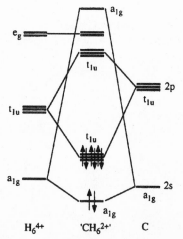

Figure 8.5
An orbital diagram for the octahedron containing an inserted small main group atom

References

1. Hoffmann, R., *Angew. Chem., Int. Ed.*, **21**, 711 (1982).
2. Ellis, J. E., *J. Chem. Educ.*, **53**, 2 (1976).
3. Wade, K., *Chem. Commun.*, 792 (1971); *Inorg. Nucl. Chem. Lett.*, **8**, 559, 563 (1972); *Adv. Inorg. Chem. Radiochem.*, **18**, 1 (1976).
4. Mingos, D. M. P., *Nature (Phys. Sci.)*, **236**, 99 (1972); *Chem. Soc. Rev.*, **15**, 31 (1986); *Accts. Chem. Res.*, **17**, 311 (1984).
5. Albright, T. A., Burdett, J. K. and Whangbo, M.-H., *Orbital Interactions in Chemistry*, John Wiley & Sons (1985).
6. Mingos, D. M. P., *Chem. Soc. Rev.*, **15**, 31 (1986).

9 What is the Metallic Bond?

We have classified molecules and solids for many years in terms of their bonding type; ionic, covalent, van der Waals and metallic. These have been very useful pigeonholes that help us considerably in terms of our expectations concerning their properties. The metallic bond however, is a category that seems out of place today. This is especially true when we consider Figure 9.1, from a study by David Pettifor[1]. He showed that by using the solid-state analog of simple Hückel theory (an orbital construct) one could readily calculate the relative energies of the fcc, hcp and bcc arrangements for the transition elements as a function of the number of electrons and accurately predict their structures. The only two errors seem to be for iron (a magnetic, or high-spin system) and manganese, although γ- Mn, a less stable allotrope at room temperature has the fcc structure and δ-Mn the bcc structure. Thus it does not seem as if we need to use a special model to study these systems at all. The 'metallic' bond looks as though it's just a more complex form of the orbital models we use for 'covalent' molecules.

You are right. Probably the term 'metallic bond' should be dropped from chemical usage. But let's take a look at some history. The metallic bond has its origin in the free-electron model of the physicist[2]. Here the electrons are effectively free to move in a box perturbed by the potentials of the nuclei located in it. The chemist in the past has, by and large, seen the large coordination numbers (twelve in the close-packed structures), despaired of writing two-center two-electron bonds to describe the interatomic interactions and been happy to have some other type of chemical bonding picture to describe this situation. Linus Pauling however, recognized[3] that the electronic picture for metals was similar in many ways to that in other chemical problems where there are not enough electron pairs to satisfy traditional bonding ideas at each center. He visualized the metal as

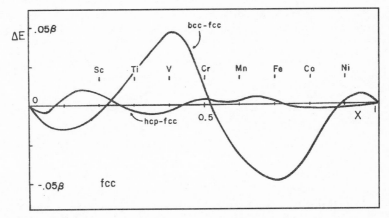

Figure 9.1

Calculated energy differences between the hcp, fcc and bcc structures of the transition elements using a d-orbital-only tight-binding model. The transition metals are described electronically by the overlap of metal d and s bands. There is roughly a population of one s electron for all of the transition metals. Thus when comparing this simple theory and observation, each metal has one 'd-electron' less than expected from its position in the Periodic Table. This is taken care of in the diagram by elemental labels

being described by a whole series of resonating valence bond structures just like that for benzene in Structure **12.3**. This Resonating Valence Bond Model has proven very useful over the years. Thus, early on in the development of theories of metallic bonding it was recognized that metals provided just a more complex aspect of existing chemical bonding ideas. Pettifor's result of Figure 9.1 underscores this and the utility of orbital pictures in general.

> Of course there are many, often unusual, materials to be added to our metallic database since Pauling's time. There are a substantial number of solids described as molecular metals held together largely by van der Waals interactions. Some of the most recent examples are the doped fullerenes that are not only metals but are sometimes superconductors at low enough temperatures. There are also metals that are generated by the doping of solids traditionally regarded as 'ionic', such as $La_{2-x}Sr_xCuO_4$. Here a superconductor (at low enough temperature) or normal metal is found for $x > 0.05$.

Yes, a variety of 'bonding types' are found as metals. As you noted earlier, the traditional spectrum of chemical bonding possibilities includes four types of 'bonds' describing ionic, covalent, van der Waals and metallic interactions. (We'll ignore hydrogen bonds in this classification even though they are of immense importance in many areas of chemistry.) The first three apply to both molecules and solids but the last is only applicable to solids. In any complete quantum mechanical calculation we would expect to find terms that describe each of these. In simpler calculations the essence of the theory would turn up in rather specific ways. For ionic interactions, we use an interatomic potential of the form $-e^2/r + A/r^n$, where the electrostatic interaction is balanced by a repulsive term. This is an approach that has been extensively used to estimate the cohesive energies of ionic solids. For van der Waals interactions, the simplest treatment uses a potential of the form $-B/r^6 + C/r^{12}$, the Lennard-Jones 6–12 potential. To estimate the strength of covalent bonding we evaluate integrals of the form $\langle \phi_1 | \mathscr{H}^{\text{eff}} | \phi_2 \rangle$. But there is no special electronic characteristic that allows us to estimate the strength of 'metallic bonding'. This is another observation that tells us that this concept does not carry the same weight as the other three.

> How are metals regarded from a physical perspective?

From the physics point of view the vital consideration for metallic conduction is the presence of empty electronic levels just above the Fermi level, for our purposes the highest occupied level in the solid. It means that an entering electron finds a low energy level in which to move. This implies that the energy band containing the conduction electrons must be only partly filled. (We discuss the formation of energy bands in general in Chapter 10.) A system with a filled band will not be a conductor. Figure 9.2 shows some possibilities. Diamond with a filled s/p band (see the discussion around Figure 3.5) and a large gap between filled and empty bands is an insulator, chromium with a partially filled s/d band, a metal. This physical requirement of a partially filled band is very important when viewing chemical aspects of this question.

The key to what is happening here lies in the concepts of delocalized and localized bonding described in Chapter 3. Metals are frequently viewed as being described by a de-localized electronic picture because they are electrical conductors. Such 'delocalization' allows electrons to move readily from one atom to another. Insulators are correspondingly described by a localized picture since the electrons do not move freely in the material.

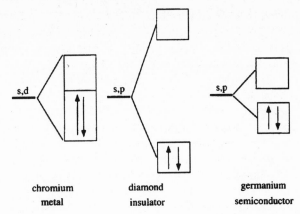

Figure 9.2
Some metals and insulators

However, remember that both localized and delocalized approaches are equally applicable for BeH$_2$ and methane as long as one is interested in collective properties. In solids the same considerations apply. Thus we can go back and forward between localized and delocalized descriptions of the electronic structure of many octet solids. These include diamond, most silicates and many of the Zintl phases*, for example, where there are the same number of pairs of electrons in bonding orbitals as there are 'bonds'. The approach is only a little more complicated than that in Structure **3.2**. However, recall that only the 'delocalized' picture is available for the π system of benzene (Structure **3.5**) since there are fewer electron pairs than bonds (here defined as two-center two-electron moieties between close contacts of pairs of atoms). In other words the option of a localized electronic description is available only for a subset of molecules and solids. (This is assuming that we use a localized description in terms of two center-two electron bonds.) However, the delocalized picture is always appropriate.

Since elemental scandium has a close-packed structure with 12 nearest-neighbors but only three valence electrons, of course the delocalized model will be the only one applicable here.

Now in solids these results may be put on a formal footing by introducing the idea of the Wannier function[4] (equation (9.1)) a route towards the generation

$$W \propto \sum_k \exp(-i\mathbf{k} \cdot \mathbf{r})\psi(k) \tag{9.1}$$

of localized functions in solids. We will not spend any time justifying the terms in this equation (you can read about it elsewhere[4]), but note that the Wannier function W represents a weighted sum (over the label k) of the functions $\psi(k)$, just as in the much simpler case of the localization of the molecular orbitals in BeH$_2$ (Figure 3.3(b)) or of methane (Structure **3.2**). The label k in its simplest form describes all the individual

---

* These are solids containing electropositive atoms in addition to main group elements from the right-hand side of the Periodic Table. The electropositive atom effectively donates an electron to the latter such that the structure found is one expected for the main group element plus these additional electrons. Thus KGe, best written K$_4$Ge$_4$, contains Ge$_4^{4-}$ tetrahedra isostructural and isoelectronic with P$_4$. CaIn$_2$ has the hexagonal diamond structure for the In framework, a structure expected for a Group 14 element. The Ca atoms sit in holes in the structure.

orbitals in an energy band as we will see in Chapter 10. Thus the sum over all k represents a sum over all the levels in the band. So, in order to generate a localized set of orbitals the band needs to be completely filled such that the sum of equation (9.1) over all filled orbitals does include those of all k. If the band is partially full of electrons then a de-localized picture only is possible and a metal will result.

This discussion then leads to one view of the 'metallic bond'. It is just a delocalized description of the electronic situation but without the option of the construction of localized Wannier functions. Since the construction of such localized functions demands the presence of a filled band of electrons, a metal is found for a partially filled energy band. So you can see that there is no special type of chemical bonding associated with the 'metallic' bond. The orbital interactions of the type used to view all sorts of molecules and solids are just as applicable for metals as implied by the results of Figure 9.1. Recent years have seen great success in the study of metals using tight-binding theory, really just LCAO ideas applied to solids as we will see in Chapter 10. Let's drop the term 'metallic bond' from use.

References

1. Pettifor, D. G., *Calphad*, **1**, 305 (1977).
2. There are discussions in many basic physics texts. See, for example Ashcroft, N. W. and Mermin, N. D., *Solid State Phys.*, Saunders (1976).
3. Pauling, L., *The Nature of the Chemical Bond*, third edition, Cornell University Press (1960).
4. See Burdett, J. K., *Chemical Bonding in Solids*, Oxford University Press (1995).

General References and Further Reading

Alonso, J. A. and March, N. H., *Electrons in Metals and Alloys*, Academic Press (1989).
Bube, R. H., *Electrons in Solids*, Academic Press (1992).
Burns, G., *Solid State Phys.*, Academic Press (1985).
Callaway, J., *Energy Band Theory*, Academic Press (1964).
Hatfield, W. E. (editor), *Molecular Metals*, Plenum (1979).
Pettifor, D. G., *Bonding and Structure of Molecules and Solids*, Oxford University Press (1995).
Sutton, A. P., *Electronic Structure of Materials*, Oxford University Press (1993).

10 Is the Electronic Description of Solids Just Like That for Large Molecules?

If I take the general Hückel result for a chain of carbon $p\pi$ orbitals given by equation (2.20) (and (10.1)) and ask what happens when the chain is of infinite length

$$E(j) = \alpha + 2\beta \cos(j\pi/N + 1) \tag{10.1}$$

I get the infinite collection of orbitals bounded by the energies of $\alpha \pm 2\beta$ shown in Figure 10.1. I might call this an energy band which has a width $W = |4\beta|$. The orbital behavior of this infinite situation is just like that in the molecules of Figure 2.1, the orbital at the very bottom of the band is in-phase (bonding) between orbitals on adjacent atoms and the orbital at the very top, out-of-phase (antibonding). It looks as though there is no real difference between the way we describe small molecules and the infinite one, namely the solid.

This is quite right; Hückel theory in the hands of the molecular chemist becomes tight-binding theory when the same LCAO ideas are applied to solids. Although your approach is quite revealing, probably the most useful way to relate the orbitals of molecules and solids[1,2] is to extend the Hückel results for small cyclic polyenes to the infinite one. First we need to visualize the infinite solid as in Structure **10.1**. Here the ends of the chain have been tied together such that the 'molecule' has no end atoms (as in the infinite solid). Now the energy levels of the cyclic polyene are given by equation (10.2) (and (13.1)) which shows that there are N energy levels labeled by the quantum number j, which takes all integral values from $0, \pm1, \pm2 \ldots \pm N/2$.

$$E_j = \alpha + 2\beta \cos(2j\pi/N) \tag{10.2}$$

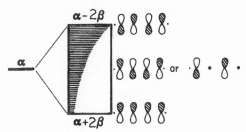

Figure 10.1
Extension of the general Hückel result for a finite chain of carbon $p\pi$ orbitals to the infinite case

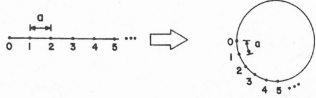

1

For the small molecule this is easy to handle but becomes clumsy for the solid since $N/2$ is now extremely large. Equation (10.2) may be rewritten for this case in a much tidier fashion as

$$E(k) = \alpha + 2\beta \cos ka \tag{10.3}$$

by defining a new index $k = 2j\pi/Na$, called the wavevector, which takes values from 0 continuously in the range $-\pi/a < k \leqslant \pi/a$. Here a is the unit cell length of Structure **10.1**. Figure 10.2 shows a very interesting way to show the transition[3] from the finite to the infinite case. In the crystalline state the space containing the levels between $-\pi/a < 0 < \pi/a$ is called the first Brillouin zone. The point $k = 0$ lies at the zone center, $k = \pm\pi/a$ is the zone edge and the energy variation with k is the dispersion of the band. The obvious point to note is that the curve of Figure 10.2(c) is just a 'smoothed out' version of Figure 10.2(b). A continuum of levels (in the limit $N \rightarrow \infty$) has replaced the discrete set of the molecular case.

This is really an elegant approach, but how else should we regard k?

We need to go back to the orbitals of the cyclic molecules to reveal another meaning of the label k. We can readily construct the energy levels of these systems, such as those in benzene and cyclobutadiene, using group theory. Quite generally, for a point group G which contains the operations g ($g \in G$), a symmetry-adapted linear combination of atomic orbitals may be generated for the kth irreducible representation using a projection operator as

$$\psi_k = \sum_{g \in G} \chi_k(g) g \phi_1 \tag{10.4}$$

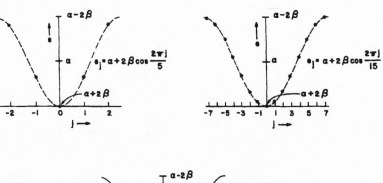

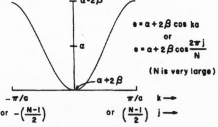

Figure 10.2
The transition from the finite to the infinite cyclic molecular case

where ϕ_1 is some member of the orbital basis and $\chi_k(g)$ is the character of the kth irreducible representation for the gth operation. We can use the C_n point groups to describe the energy levels of these cyclic molecules even though their real point symmetry is higher. The C_n point groups have in fact a special characteristic associated with their character tables. There is one totally symmetric (and non-degenerate) representation as in all groups, and for $n = $ odd all of the other irreducible representations occur in pairs. (Table 10.1 shows the character table for the C_5 group.) Importantly, for a cyclic $(CH)_n$ molecule there will be $n\pi$ molecular orbitals of course, one belonging to *each* irreducible representation of the cyclic group of order n, C_n. Now the labels j in Figure 10.2 may be identified in addition with a group-theoretical label from the molecular point group.

The wavefunctions of these cyclic molecules may be written down by using equation (10.4) and the characters of the C_n group in question (equation (10.5)). This looks complex but the exponential is simply the character $\chi_j(C_n^{p-1})$ of the cyclic group of order n as you may readily check for the case of $n = 5$ shown in Table 10.1. The prefactor of $n^{-1/2}$ is a (Hückel) normalization constant.

$$\psi_j = n^{-1/2} \sum_{p=1}^{n} e^{2\pi i j(p-1)/n} C^{p-1} \phi_1$$
$$= n^{-1/2} \sum_{p=1}^{n} e^{2\pi i j(p-1)/n} \phi_p \tag{10.5}$$

There is an obvious reason for doing this. We can get the infinite case just by the substitution of $k = 2j\pi/na$ in equation (10.5) as

$$\psi(k) = n^{-1/2} \sum_{p=1}^{n} e^{ik(p-1)a} \phi_p \tag{10.6}$$

Let's now look at the translation group T for the solid. This is an infinite group made up of all translations, t ($t \in$ T). In the solid we can use the projection operator as before, and write

$$\psi(k) = n^{-1/2} \sum_{t \in T} \chi_k(t) t \phi_1 \tag{10.7}$$

If R_t is the distance along the chain that the translation operation t moves ϕ_1 and ϕ_t is an orbital translationally equivalent to ϕ_1, but located at R_t, then $t\phi_1 = \phi_t$.

$$\psi(k) = n^{-1/2} \sum_{t \in T} \chi_k(t) \phi_t \tag{10.8}$$

Table 10.1   The Character Table for the C_5 group

C_5	E	C_5	C_5^2	C_5^3	C_5^4	
A	1	1	1	1	1	z, R_z
E_1	$\begin{cases} 1 \\ 1 \end{cases}$	$\begin{matrix} \varepsilon \\ \varepsilon^* \end{matrix}$	$\begin{matrix} \varepsilon^2 \\ \varepsilon^{*2} \end{matrix}$	$\begin{matrix} \varepsilon^{*2} \\ \varepsilon^2 \end{matrix}$	$\begin{matrix} \varepsilon^* \\ \varepsilon \end{matrix} \Big\}$	$(x, y)(R_x, R_y)$
E_2	$\begin{cases} 1 \\ 1 \end{cases}$	$\begin{matrix} \varepsilon^2 \\ \varepsilon^{*2} \end{matrix}$	$\begin{matrix} \varepsilon^* \\ \varepsilon \end{matrix}$	$\begin{matrix} \varepsilon \\ \varepsilon^* \end{matrix}$	$\begin{matrix} \varepsilon^{*2} \\ \varepsilon^2 \end{matrix} \Big\}$	

$\varepsilon = 2\pi i/5$

Importantly the characters of the translation group have the same simple exponential form found for the cyclic groups. In fact the infinite translation group is isomorphic to the infinite cyclic group of order n. Thus

$$\psi(k) = n^{-1/2} \sum_{t \in T} e^{ikR_t} \phi_t \qquad (10.9)$$

This is identical to equation (10.6) but with a trivial change of labels.

When we write $R_t = (p-1)a$ it is clear that equations (10.6) and (10.9) are identical. So the label k is not only a quantum number, like j in the cyclic molecule, but is also a symmetry label as in the molecule too. From Figure 10.1 it is, interestingly, a node counter as well.

Yes, this connection between molecules and solids in terms of symmetry is an interesting one. You pointed out the nodal aspects of the orbitals of the energy band earlier and certainly as k increases, the energy increases and the number of nodes increases as we can see in Figure 10.1. The levels are bonding at the bottom and antibonding at the top of the energy band.

What do the wavefunctions look like?

This comes directly from equation (10.9) and is just as anticipated from direct extrapolation of the behavior of molecules. The e^{ikR} term tells us that at $k=0$ (bottom of the band) the wavefunction is in-phase (a phase factor of $+1$) from cell to cell, but that at $k=\pi/a$ (top of the band) it is out of phase (a phase factor of -1). This result (Structure **10.2**) is exactly in accord with what is expected by extrapolation of the molecular case.

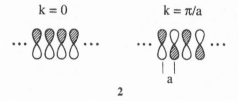

2

Is there a way we can show the bonding and antibonding properties of a band from experimental data?

There is a nice way of probing the antibonding nature of the orbitals at the top of the energy band that comes from the structural changes found by doping materials such as $K_2Pt(CN)_4$. Just as we were fortunate in the infinite polyene to be able to use a single orbital per center, here the energy band is generated by interaction of the z^2 orbitals of the square planar units in Structure **10.3**. The planes stack[2] one above the next as in Structure **10.4** to produce a chain with the z^2 orbitals full as expected for a low-spin d^8 (Pt(II)) configuration. (The bandwidth is 4β, just as in the polyacetylene case, but here β is the interaction integral linking adjacent metal z^2 orbitals rather than adjacent $p\pi$ orbitals.) With a full band (Structure **10.5a**), the compound $K_2[Pt(CN)_4]$ is an insulator $(5 \times 10^{-7} \ \Omega^{-1} \ cm^{-1})$ and the Pt–Pt distances along the chain are 3.48 Å. However, the

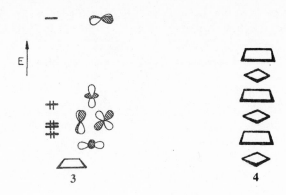

3   4

salt can be co-crystallized with elemental bromine. This results in oxidation of the chain to give the non-stoichiometric material $K_2(Pt(CN)_4)Br_\delta.3H_2O$. Importantly the system now has to be formulated as $Pt(CN)_4^{\delta-2}$. Thus electron density has been removed from the z^2 band (Structure 10.5b). This material is now metallic and, since electrons have been removed from the very top of the band where maximum antibonding interactions are found, it has a much shorter Pt–Pt distance (2.88 Å for $\delta = 0.3$).

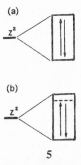

5

Do similar considerations apply to 'real' solids, namely those with more than one orbital per center, and with atoms of different types located in more than one dimension?

The answer is yes, but it is more complicated of course. Whereas in the molecular case (equation (2.13)) we solved the secular determinant $|H - SE| = 0$, now we need to solve the equation $|H(k) - S(k)E| = 0$. We have written k as a vector since it will in general describe the behavior of k in a three-dimensional space. However, just to show how this works let's look at a two-atom cell (Structure 10.6). It's a trivially simple example but shows many of the features found in 'real' systems. Since there are two orbitals in the cell this behaves as a two-atom molecule. There will be two bands whose energies will be given by the roots of the determinantal equation we have just described. First let's write down the wavefunction using equation (10.9) for the orbitals at the left and right-hand side of the cell. These are just

$$\psi_1(k) = n^{-1/2}(\ldots\phi_1 e^{-ika'} + \phi_2 + \phi_3 e^{ika'}\ldots)$$
$$\psi_2(k) = n^{-1/2}(\ldots\phi_4 e^{-ika'/2} + \phi_5 e^{ika'/2} + \phi_6 e^{3ika'/2})$$

(10.10)

In order to set up the 2×2 secular determinant we need to evaluate the H_{ij}. The H_{ii} are independent of k since there is no interaction (in the Hückel approximation) between ϕ_1 in one cell (for example) and ϕ_2 or ϕ_3 in adjacent cells. So

$$H_{11} = \langle\psi_1(k)|\mathcal{H}^{eff}|\psi_1(k)\rangle = n^{-1/2}n^{-1/2}(n\alpha) = \alpha$$

(10.11)

and analogously

$$H_{22} = \langle\psi_2(k)|\mathcal{H}^{eff}|\psi_2(k)\rangle = n^{-1/2}n^{-1/2}(n\alpha) = \alpha$$

(10.12)

The off-diagonal element H_{12}, however, contains interactions between (say) ϕ_2 and ϕ_4 and ϕ_5. Within the Hückel approximation it depends on β and $\mathbf{k}$. It is readily evaluated as

$$H_{12} = \langle\psi_1(k)|\mathcal{H}^{eff}|\psi_2(k)\rangle = n^{-1/2}n^{-1/2}n(e^{ika'/2} + e^{-ika'/2})\beta$$
$$= 2\beta\cos ka'/2.$$

(10.13)

So the secular determinant becomes

$$\begin{vmatrix} \alpha - E & 2\beta\cos ka'/2 \\ 2\beta\cos ka'/2 & \alpha - E \end{vmatrix} = 0$$

(10.14)

with roots $E = \alpha \pm 2\beta\cos ka'/2$. This result is shown in Figure 10.3. Now there are two orbitals for each value of $\mathbf{k}$. You can see that the $E(\mathbf{k})$ diagram of the two-orbital cell is that

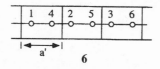

of the one-orbital cell (Figure 10.2(c)) with the levels folded back along $k = \pi/2a$, shown pictorially[2,3] in Figure 10.4. In 'real' systems we just have a larger determinant (more orbitals), non-zero overlap integrals and diagonal terms that contain some of these long-range interactions. Usually for most systems of interest this has to be solved numerically.

Let's take a complex system with many atoms and orbitals in the unit cell of different types. How do we display the results of a calculation? Surely we don't do a calculation at every value of $\mathbf{k}$. This would lead to an infinite number of orbitals.

No, we don't calculate an infinite number of energies. We actually calculate a finite set determined by a special representative set of $\mathbf{k}$ points[1,2]. So for the one-dimensional system we would calculate a set of points evenly spaced in this 'k-space' of Figure 10.2(c). The results are usually displayed in terms of a density of states diagram (Figure 10.5). Here the abscissa tells us the number of energy levels within a given energy interval. Invariably we smooth the histogram produced by such counting to produce an aesthetically more

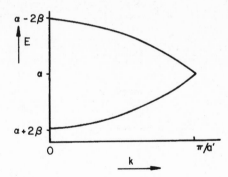

Figure 10.3
The E(k) diagram for the two orbital cell

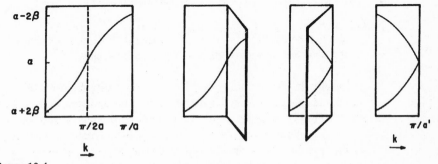

Figure 10.4
A picture showing how the orbitals of the two-orbital cell are just those of the one-orbital cell but folded back along $k = \pi/2a$

appealing picture. Figure 10.6 shows the corresponding density of states picture for the π orbitals of benzene. Of course, in molecular chemistry traditionally we do not use this form of the presentation of the data. Figure 10.7 shows the calculated density of states for TiO_2

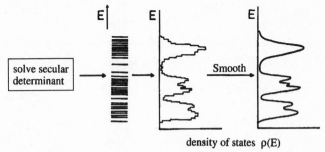

density of states $\rho(E)$

Figure 10.5
The density of states diagram

in the rutile structure. Notice the energy bands that develop in the regions appropriate for oxygen 2p and Ti 3d orbitals.

Clearly the analog with the cyclic molecules is a useful one but can it go further? For example cyclobutadiene is Jahn–Teller unstable and undergoes a dimerization, the benzene dication

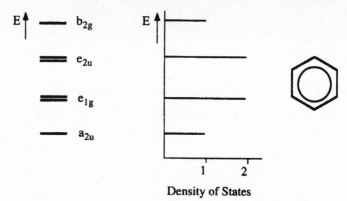

Figure 10.6
The density of states description of the π orbitals of benzene

should distort to form two trimers if it remained planar, and so on. Presumably, if the analogy is a good one similar distortions should happen in solids.

Exactly. In the one-dimensional chains (polyacetylene for example) similar distortions occur, the energetics being a balance between the demands of the σ and π manifolds just as described in Chapter 12. Figure 10.8 shows what happens energetically to cyclobutadiene and to the infinite polymer if adjacent bonds are alternately shortened and lengthened. This is the way to relieve the Jahn–Teller instability in cyclobutadiene but notice how the highest occupied orbitals in the infinite polymer are stabilized too. (The top and bottom of the band have energies of $\alpha \pm 2\beta$ (Figure 10.1) just as in the molecule.) Such a Jahn–Teller distortion in solids is known (after Rudolf Peierls) as a Peierls distortion[4]. In general, a

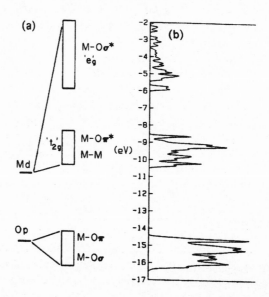

Figure 10.7
(a) Expected location of the energy bands of TiO_2 in the rutile structure; (b) calculated electronic density of states

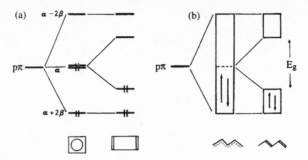

Figure 10.8
*(a) The Jahn–Teller instability in cyclobutadiene; (b) the Peierls distortion in the infinite polymer.
Both processes involve the alternate shortening and lengthening of adjacent bonds*

one-third or two-thirds full band will distort to form trimers and a quarter, or three quarters full band will distort to form tetramers (Structure **10.7**). Some examples of Peierls distortions are shown in Table 10.2. Notice the nice example of the hydrogen atom chain.

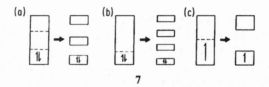

7

Electronically it looks very similar to polyacetylene, the short/long bond alternation leading to H_2 molecules. There are some comments of yours concerning oligomers of hydrogen in Chapter 12 and we'll come back to hydrogen later.

> There is a particularly interesting point concerning Structure **10.7** and that is associated with your quarter-filled band examples. If the electrons are all paired up then the band is one-quarter filled with paired electrons and a tetramer is found, but if with the same electron count it is half filled with unpaired electrons, then a dimerization is found.

Yes, it is an interesting feature. The question as to which is found for a particular case is a difficult one to answer though. Electronically it is somewhat similar to the case of the Jahn–Teller instability of octahedral d^8 metal complexes described in Chapter 14. There is a balance between the strength of orbital interactions that determines the bandwidth and the electron–electron interactions that destabilize paired electrons.

In more than one dimension things get a bit more complicated. To put it simply, what might be a stabilization in one direction may be unfavorable electronically in another. So TaS_2, on cooling undergoes a distortion that leads to a reduction in its metallic behavior but does not lead to an insulator with a band gap. However, some systems may be described electronically as collections of independent one-dimensional systems[1]. For example Figure 10.9 shows the simple cubic structure and two of the many possible structures that may be obtained by alternately breaking bonds along each of the three directions. These are the black phosphorus and arsenic structures. Figure 10.10 shows a simple band picture for these Group 15 solids. Let us assume that two electrons reside in a valence s orbital and that the remaining three occupy the bands generated by the p orbitals.

Table 10.2 Peierls distortions in linear chains

(1) Polyacetylene	Doped polyacetylene
bond alternation, semiconductor	metal
(2) NbI_4 chain	NbI_4 under pressure
pairing up of metal atoms	metal atoms equidistant, metal
(3) VO_2 chain (rutile structure)	VO_2 at lower
metal atoms equidistant	temperatures—pairing up of metal atoms
(4) Elemental hydrogen	High pressure-metallic behaviour
H–H dimers	...H–H–H–H... chains[a]
(5) $BaVS_3$ [VS_3 chain]	Metal–insulator transition
metallic at room temperature	on cooling, but a magnetic insulator rather than diamagnetic
(6) (TTF) Br	(TTF) $Br_{0.7}$
$(TTF)_2^{+2}$ dimers insulator	metallic conduction cf. polyacetylene

[a] Actually some close-packed structure in practice.

If we imagine that each p_j ($j = x, y, z$) orbital pointing along the jth direction only participates in σ interactions in that direction then the result is three independent, degenerate bands. Counting electrons fills them each half full. (Write the electron configuration of the atom as $s^2 p_x^1 p_y^1 p_z^1$ to see this.) We might then expect a Peierls dimerization along each direction, which is what is seen exactly in the structures of Figure 10.9. Some derivative structures (NaCl, SnSe, SnS) are labelled too.

We talked about the metallic bond in Chapter 9. Clearly a metal is just a solid with a partially filled energy band. But partially filled bands are susceptible to a Peierls distortion which in principle leads to an insulator. Presumably in metals this distortion is frustrated by the underlying structure. There must be some nice examples.

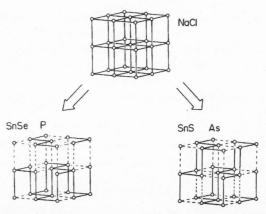

Figure 10.9
The simple cubic structure and the black phosphorus and arsenic structures that may be obtained by alternately breaking bonds along each of the three directions

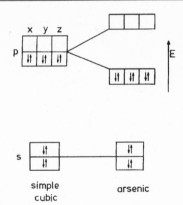

Figure 10.10
A simple band picture for the Group 15 solids of Figure 10.9

There is a wealth of interesting metal–insulator problems[5]. They bear on your comment concerning the frustrated Peierls distortion and the balance between the geometrical preferences of valence and deeper-lying orbitals noted in Chapter 12. If one can switch off the Peierls distortion in a solid then one should be able to reverse the process of Figure 10.8(b). This can be done in several ways. One is to apply pressure and another to dope the material with electrons away from the optimal Peierls electron count.

At high pressures with a reduction in interatomic separation the structural preferences of solids are strongly controlled by the repulsive part of the interatomic potential. The result is that all of the bond lengths want to be equal. We can see this by looking at the variation in the repulsive energy, described by a potential of the form B/r^n. The difference in repulsive energy for the two examples in Structure **10.8** is proportional to $(2a)^{-n}x^2$, the repulsion being less fierce for the symmetrical structure. Thus under high pressure when the energy of the system is dominated by the repulsive part of the potential, the most stable structure will be the one where all bond lengths are equal. The simplest problem, and one very similar (Table 10.2) to that of the polyacetylene chain, is the hypothetical one of a linear chain of hydrogen atoms, Structure **10.9**. This should undergo a Peierls distortion, and does, the resulting dimers in accord with traditional ideas concerning bonding in this molecule. However, it is virtually certain that application of high pressure ($> \sim 4$ Mbar) reverses the process (Structures **10.9–10.10**) and that a metal is produced. The high pressure required is directly connected with our discussion in Chapter 12. Such a structure (or rather its three-dimensional analog) with its metallic properties is of interest to

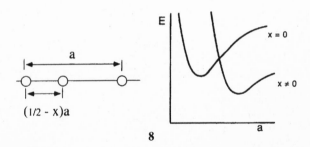

8

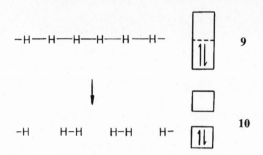

geophysicists, since a part of the bulk of the Jovian planets has been suggested to be made from such material.

Doping the energy bands with electrons or holes (i.e., removing electrons) may be done either 'formally' or 'informally'. In the first case electrons are either added to the next highest energy band (the conduction band) as in Na_3C_{60} or removed from the highest filled band (the valence band) as in $(TTF)Br_{0.7}$, as we see below, by the addition of a reducing or oxidizing agent respectively. Metals result in both cases. Informally the valence band may be depleted and the conduction band populated by heating. In the case of VO_2 there is a transition from a Peierls-distorted insulator to undistorted metal at 340 K.

How do the 'molecular metals', such as the fullerenes, fit into the picture?

These are a particularly fascinating class of solids since they often contain no 'metal atoms' at all yet are frequently metallic[6]. $(TTF)Br_{0.7}$, mentioned above, and many of its derivatives have been known for many years but perhaps the most recent examples are the systems you mentioned, the doped fullerenes[7,8] M_3C_{60}. They contain close-packed fullerene molecules with electropositive metal atoms (Na, etc.,) in the interstices in this structure. Although these solids do contain metal atoms they are there as M^+ ions and the fullerene unit is correspondingly reduced. As we can see in Structure **10.11** the parent

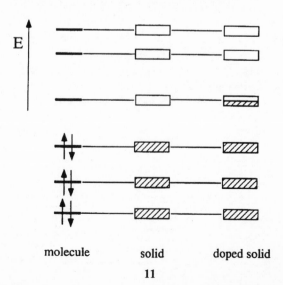

11

molecular solid is insulating, but on reduction a partially filled band results and at the C_{60}^{3-} stoichiometry a metal is generated. Other materials of the same type are characterized structurally by the stacking of planar molecules such as those of Structure **10.12** in the manner shown in Structure **10.13**. In the pure state, the solid is not a conductor of electricity; these solids look just like many other organic solids. However, this time on oxidation, often achieved by co-crystallization with halogen, a conducting material can result. A series of solids[1] involving TTF (tetrathiofulvalene) is shown in Figure 10.11. The undoped material is an insulator and a typical organic solid, but $(TTF)Br_{0.7}$ is a metal since it now has a partially filled band. With a half-filled band as in $(TTF)Br_{1.0}$, a Peierls distortion occurs leading to dimerization $((TTF)_2^{2+}$ pairs), in a way analogous to the generation of H_2 molecules from the one-dimensional chain for hydrogen. If the band is completely empty, individual $(TTF)^{2+}$ ions are found and an insulator results.

Although the distances between the units of such species, C_{60}^{3-} or TTF, are reasonably large, they are in fact short enough to allow significant orbital overlap between adjacent molecules and the formation of an energy band within which conduction may occur. There is no 'metallic' bonding between the units.

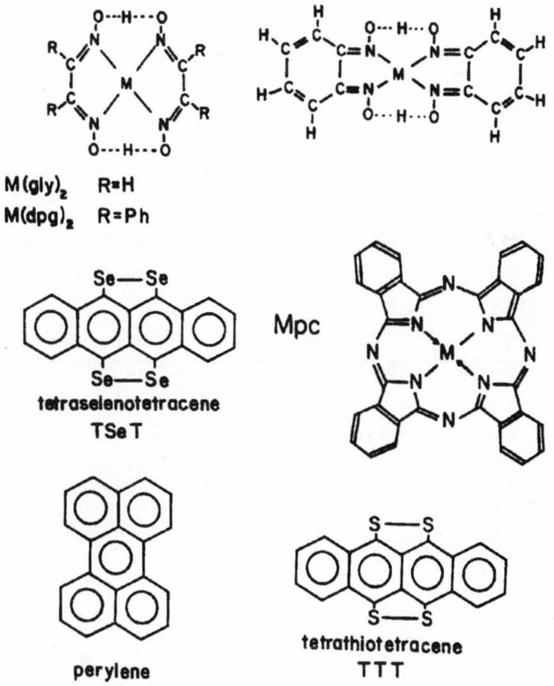

M(gly)$_2$ R=H
M(dpg)$_2$ R=Ph

tetraselenotetracene
TSeT

Mpc

perylene

tetrathiotetracene
TTT

12

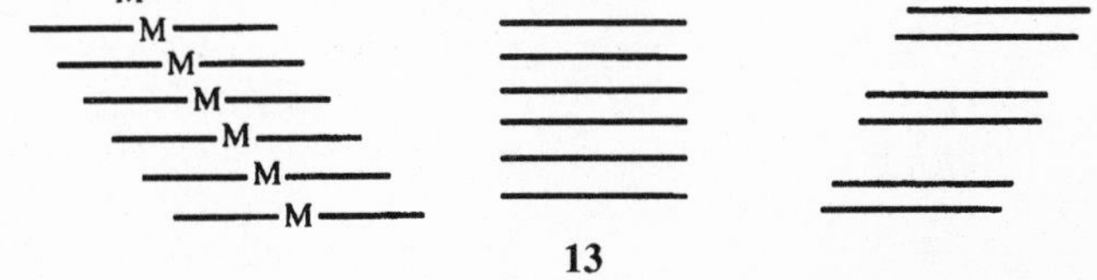

13

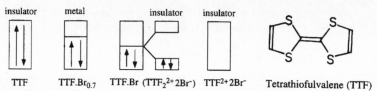

Figure 10.11
Some solids involving TTF (tetrathiofulvalene)

Clearly considerations such as these are very important in understanding solid state structures. But what about energetics? The bond energies of molecules, related to the bond order in the simplest sense, show a maximum when all of the bonding orbitals are filled. So N_2 has the largest bond dissociation energy of any diatomic when three bonding orbitals are filled with electrons. Are there similar simple concepts for solids? One would predict maximum stabilization for the half-filled band using the same type of idea (Structure **10.14**).

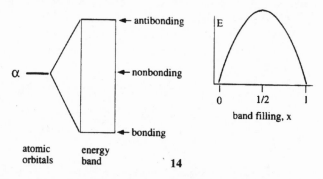

Yes, there are some quite analogous ideas developed by Jacques Friedel[9]. Figure 10.12 shows the variation in the cohesive energy for solids of the main group elements (the sp elements) and the transition metals[1] (the d elements). Notice the inverted parabolic dependence on electron count, the plots peaking in the middle of the series of either the half-filled $(s + p)$ band for the main group elements or the corresponding half-filled $s + d$ band for the transition metals. We'll ignore the bumpy nature of the plot for the first transition metal series. As you noted there is a similar dependence on bond order or vibrational force constant (shown in Structure **10.15**) for the diatomic molecules of the first

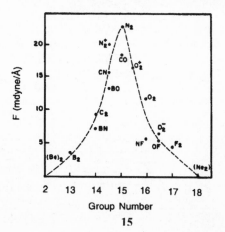

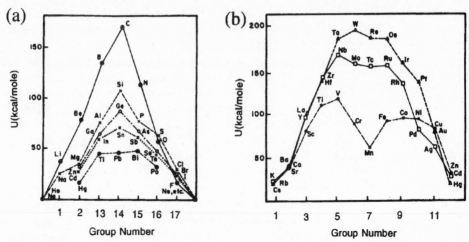

Figure 10.12
The variation in the cohesive energy of solids of the main group elements and the transition metals

row elements. The explanation for both sets of plots is the same. As one fills up bonding orbitals (those below α in your picture of Structure **10.14**) the material, whether molecule or solid, is stabilized. The maximum stability is found at the half-filled point. More electrons populate anti-bonding orbitals and now the cohesive energy or bond order decreases. It's really rather simple.

It doesn't seem to matter what the crystal structure or geometry is of the material under consideration.

No, by and large it doesn't. In general the cohesive energy of the solid is much larger than the energy associated with any structural variation. So while the cohesive energy of the anatase and rutile polymorphs of TiO_2 are around 940 kJ/mol, the energy difference between them is only 8 kJ/mol. This small difference often makes it difficult to identify the lowest energy polymorph by calculation since the energy differences between structures may be quite small.

If I were to perform a calculation, along the lines you have described, for ruby, I would probably find that the system is predicted to be metallic. A partially occupied band would result from the partially filled d levels of Cr(III). This is certainly not the case in practice. What is missing from the theory?

What we have described in this chapter so far is an electronic model, derived as you suggested initially, from the delocalized picture used in the Hückel theory for conjugated polyenes. Now as we emphasized in Chapter 5 this model will only be appropriate if the ratio of β/U (U is the on-site repulsion between two electrons on the same atom) is large. If U is large, as it is for many first-row transition metals (see Chapter 7) then this delocalized picture is inappropriate and we need to write a wavefunction that takes into account the correlation between the electrons. We know this is very important here. The electronic spectrum of ruby shows a set of transitions similar to those seen in $Cr(H_2O)_6^{3+}$, i.e., associated with a localized electronic description at chromium. The proper inclusion of this effect into the electronic structure calculation has been a challenge for solid-state theorists for a long time.

There are, though, ways to show the commonality of the different effects which lead to insulators. Figure 10.13 shows three mechanisms, each derived from the half-filled metallic band. (There are more complicated variants.) As we have shown above, a Peierls distortion or some other geometric feature of the structure (as in diamond) can open a gap (Figure 10.13(b)). Similarly, substitution of atoms of different electronegativity will lead to the same result. Thus whereas graphite is a metal and black, the isoelectronic and isostructural boron nitride is an insulator and colorless. In Figure 10.13(c) the empty band (energetically located around H_{ii} for boron 2p) will be the 'boron' band (although heavily mixed with nitrogen character) and the filled band (located around H_{ii} for nitrogen 2p) the converse, the 'nitrogen' band. Figure 10.13(d) shows an insulating state generated as a result of strong Coulombic interactions (U) between the electrons as in ruby. It has interesting connections with Figure 10.13(c). The bands are effective one-electron ones and contain the effects of such electrostatic interactions. Imagine a chain of identical atoms with a band half-filled with electrons. In the metallic state Figure 10.13(a) the electrons are free to travel throughout the solid (a delocalized wavefunction) with a probability of both being located on the same atom. As a result they experience a Coulombic repulsion. See the discussion around equation (6.2). Overall this is outweighed by the lowing of the energy due to the formation of the energy band. In the insulating state of Figure 10.13(d) we can imagine that by ensuring that no two electrons share the same site, the effective H_{ii} on this atom becomes more negative. Its adjacent atom will now have to carry two electrons and its H_{ii} will be destabilized. In other words its effective H_{ii} will become less negative. The gap between the two bands of Figure 10.13(d) depends on U. Thus the electronic picture has strong similarities to the solid containing atoms of different electronegativity. Here the different H_{ii}'s come about via Coulombic terms. In this picture we have forced the electrons in the insulator to be correlated. The gap between filled and empty bands is often called a correlation gap and its numerical calculation is quite difficult.

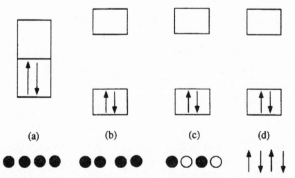

Figure 10.13
Metals and insulators: (a) a half-filled band leading to a metal; (b) an insulator, either generated from (a) by a Peierls distortion (e.g., polyacetylene) or arising as a result of the geometrical structure of the solid in another way (diamond); (c) an insulator arising from a metallic parent by the chemical substitution of atoms in the solid (e.g., graphite → boron nitride); (d) an insulator arising via strong Coulombic repulsion of pairs of electrons at each site

References

1. Burdett, J. K., *Chemical Bonding in Solids*, Oxford University Press (1995).
2. Hoffmann, R., *Solids and Surfaces*, VCH Publishers (1988).
3. Gerstein, B. C., *J. Chem. Educ.*, **50**, 316 (1973).
4. Peierls, R. E., *Quantum Theory of Solids*, Oxford University Press (1955).
5. Mott, N. F., *Metal–insulator Transitions*, Taylor and Francis (1974).
6. Hatfield, W. E. (editor), *Molecular Metals*, Plenum (1979).
7. Fischer, J. E., Heiney, P. A. and Smith, A. B., *Accts. Chem. Res.*, **25**, 112 (1992).
8. Haddon. R. C., *Accts. Chem. Res.*, **25**, 127 (1992).
9. Friedel, J. in *The Physics of Metals*, vol. 1., *Electrons*, Ziman, J. M. (editor), Cambridge University Press (1969).

General References and Further Reading

Alonso, J. A. and March, N. H., *Electrons in Metals and Alloys*, Academic Press (1989).
Bube, R. H., *Electrons in Solids*, Academic Press (1992).
Burns, G., *Solid State Physics*, Academic Press (1985).
Callaway, J., *Energy Band Theory*, Academic Press (1964).
Cotton, F. A., *Chemical Applications of Group Theory*, Third Edition, John Wiley & Sons (1990).
Pettifor, D. G., *Bonding and Structure of Molecules and Solids*, Oxford University Press (1995).
Phillips, J. C., *Bonds and Bands in Semiconductors*, John Wiley & Sons (1973).
Slater, J. C., *Quantum Theory of Molecules and Solids*, volumes 1, 2, McGraw-Hill (1963).
Sutton, A. P., *Electronic Structure of Materials*, Oxford University Press (1993).

11 How Should you Count Electrons in 'Electron-deficient' Molecules?

One class of molecules that has long provided a challenge for the chemist trying to understand their structures, is that set of molecules termed 'electron deficient', i.e., those where there appear not to be sufficient electrons to form two-center two-electron bonds between each pair of close contacts. Perhaps the boranes, typified by the smallest known member, $B_6H_6^{2-}$ (Structure **11.1**) and organometallic clusters containing metal–metal bonds such as in $Ru_6(CO)_{18}H_2$ (Structure **11.2**), are the best known. In the former with 6 B–H bonds (12 electrons) and 12 B–B bonds (24 electrons) we would expect to find a total of 36 valence electrons. However, I only find 26 if I count up the valence electrons of all the atoms, i.e., the molecule appears to be 10 electrons short. We get an interesting result if we count metal electrons for Structure **11.2**. At each metal center there are 6 electrons from the CO groups, 8 from each Ru, plus 4 from the Ru–Ru 'bonds' leading to a total of 18. This is just what we need. However, we've forgotten that there are 2 electrons from the hydrogen atoms to be shared over the 6 metal atoms. Thus the eighteen-electron rule doesn't quite seem to work. There is another interesting aspect here. The $Ru(CO)_3$ fragment is isolobal (Chapter 8) with BH, and if the two hydrogen atoms can be regarded as each contributing one electron to the cluster then $Ru_6(CO)_{18}H_2$ is isoelectronic with $(BH)_6^{2-} = B_6H_6^{2-}$. This connection is a little suspicious and surely indicates something deeper. Just how do we count electrons in species such as these?

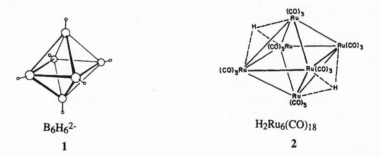

$B_6H_6^{2-}$

1

$H_2Ru_6(CO)_{18}$

2

Robert Mulliken said at one time, referring to the term 'electron deficient' molecules, that there was nothing deficient about the molecules, just the theory. Since then Kenneth Wade generated[1] a set of rules with which are can properly count electrons in these systems (Wade's rules) and Michael Mingos[2] and Anthony Stone[3] devised orbital explanations to underpin them. The compounds in question are of the type where the atoms are arranged at some, or all of the vertices of a deltahedron (a polyhedron where each face is a triangle or Δ) to give what is called either a cage or cluster compound (Figure 11.1). These clusters are formed from a variety of atomic components. They include the boranes, carboranes, and molecules containing other main group atoms, transition metal cluster compounds (Structure **11.3**), usually with cyclopentadienyl or carbonyl groups coordinated to the transition metal, compounds containing mixtures of the two building blocks, the metallacarboranes (Structure **11.4**) and molecules with a small

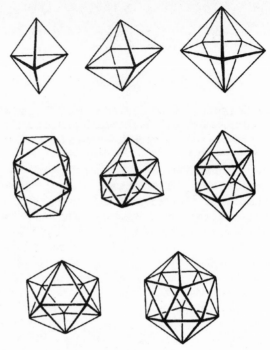

Figure 11.1
Deltahedra found for many cage and cluster molecules. They are described as the trigonal bi-pyramid, octahedron, pentagonal bipyramid, dodecahedron, tri-capped trigonal prism, bi-capped Archimedean antiprism, octadecahedron, and icosahedron

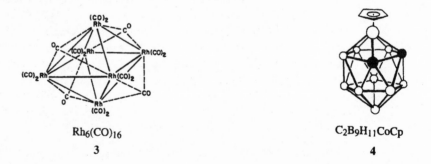

$Rh_6(CO)_{16}$

3

$C_2B_9H_{11}CoCp$

4

atom inside the cluster (Structure **11.5**) or hydrogen atoms bridging edges of the unit (Structure **11.2**). The deltahedron may be either complete (a *closo* species) as in Structures **11.1**– **11.5**, or may have one or two missing vertices (a *nido* or *arachno* species respectively) as in Structures **11.6** and **11.7**.

I suppose one should be quite general in listing a particular compound as a cluster or cage compound. So the carbocations (Structures **8.28** and **8.29**), although nicely described in terms of the isolobal analogy are nido octahedra and pentagonal bipyramids respectively.

Yes, this broader view of structure is very important. One can derive Wade's rules in various ways[4], by looking at the orbitals of some of the earlier members of the set of

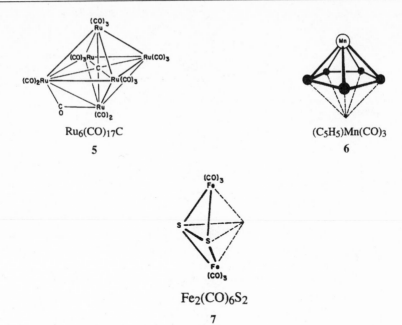

Ru₆(CO)₁₇C

5

(C₅H₅)Mn(CO)₃

6

Fe₂(CO)₆S₂

7

deltahedra shown in Figure 11.1, and trying to generalize, or by using the very powerful and general method devised by Stone[3], but for our purpose here the following is all that will be needed. We'll consider the orbital picture for main group molecules typified by $B_6H_6^{2-}$, and use the isolobal analogy to extend the results to transition-metal containing systems. At each boron atom we may envisage two sp hybrid-like orbitals, one pointing outward and coordinated to a hydrogen atom (Structure **11.8**), and the other (Structure **11.9**) which points towards the interior of the deltahedron. There are then left over two tangential p orbitals (Structure **11.10**). The outward-pointing hybrid orbital is used to attach a hydrogen atom to the boron via a simple two-center two-electron bond, and is not used in skeletal bonding. The very general orbital rule that we use is that for the *m*-vertex deltahedron. Of the *m* inward pointing radial orbitals, only one, the totally symmetric combination (Structure **11.11**), shown here for the tetrahedron, lies low enough to be occupied, and that of the 2*m* tangential orbitals, only *m* lie low enough (bonding enough) to be occupied.

This implies that such *m*-vertex molecules require a total of $m+1$ pairs of electrons for stability.

Yes, this is Wade's rule.

8

9

10   11

It looks like the three-dimensional analog of Hückel's rule for cyclic organic systems.

Indeed it is. In both cases all bonding and non-bonding orbitals need to be filled for stability. Thus for your example of $B_6H_6^{2-}$, there are $2(m+1) = 14$ electrons needed for skeletal bonding and 12 electrons for the six exo B—H bonds, leading to the 26 electrons you counted so accurately. As we see below, the same rule holds for *closo*, *nido* and *arachno* clusters. Thus the counting scheme is based on an electronic model where the electrons are delocalized all over the cluster.

It is easy to evaluate the figures for other main group units. If a BH fragment contributes two electrons (Structure **11.12**) then a CH fragment contributes three electrons (Structure **11.13**) and a sulfur atom (Structure **11.14**) four electrons. Note that we have to be careful in the last example to exclude the two electrons involved in the lone pair from the skeletal electron count. The numbers of electrons contributed to skeletal bonding by main group and transition metal-containing units are given in Tables 11.1 and 11.2, those for the latter obtained from the isolobal analogy and the values from Table 11.1. Since CH, $Co(CO)_3$, and NiCp are isolobal fragments, each contributes three electrons to skeletal bonding. Similarly BH, $Fe(CO)_3$, and CoCp are isolobal fragments and each contributes two electrons to skeletal bonding. Since $Ru_6(CO)_{18}H_2$ is isoelectronic to $(BH)_6^{2-} = B_6H_6^{2-}$, it has the same number of skeletal electrons (assuming the hydrogen atoms each contribute one electron to skeletal bonding) and the geometry of the heavy atom unit is an octahedron.

How general is the approach?

The rules are of extraordinary generality. A whole series of $B_nH_n^{2-}$ ions is known, although the smallest member with $n = 5$ has not yet been made. ($B_3C_2H_5$ is known

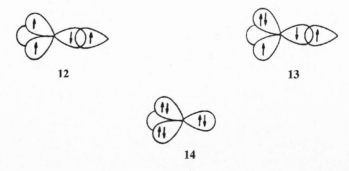

12   13

14

Table 11.1 The number of skeletal electrons contributed by main group cluster units

A	A	AH, AX[a]	AL[b]
Li, Na		0	1
Be, Mg, Zn, Cd, Hg	0	1	2
B, Al, Ga, In, Tl	1	2	3
C, Si, Ge, Sn, Pb	2	3	4
N, P, As, Sb, Bi	3	4	5
O, S, Se, Te	4	5	
F, Cl, Br, I	5		

[a] X = one electron ligand, for example, halogen.
[b] L = two electron ligand, for example, NH_3, THF.

Table 11.2 The number of skeletal electrons contributed by transition metal cluster units

M	MCp	$M(CO)_3$
Cr, Mo, W	−1	0
Mn, Tc, Re	0	1
Fe, Ru, Os	1	2
Co, Rh, Ir	2	3
Ni, Pd, Pt	3	4

however.) They have $2n + 2$ skeletal electrons and are therefore just *closo* deltahedra. One molecule of the series is particularly interesting in a historical context. This is the icosahedral molecule $B_{12}H_{12}^{2-}$ (Structure **11.15**). In 1954 molecular orbital ideas using the delocalized picture led[5] to the prediction of the −2 charge for this species many years before the molecule was actually synthesized. The authors however missed the general counting rule for these species which Wade found many years later.

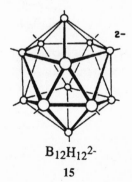

$B_{12}H_{12}^{2-}$

15

We can generalize the rules. For a carborane with the general formula $[(BH)_a(CH)_b]^{-d}$, the number of skeletal electron pairs is equal to $(1/2)(2a + 3b + d)$. By analogy with the main group systems, for a $[M_a(CO)_bCp_c]^{-d}$ cluster, the total number of skeletal electron pairs is $(1/2)[\Sigma v - 12a + 2b + 5c + d]$. Here Σv is the total number of valence electrons associated with all of the metal atoms of the cluster.

Do the rules hold just as well for clusters and cages where not all the atoms of the deltahedron are present?

Perhaps surprisingly, very similar ideas apply to the structures of *nido* and *arachno* cages. These are units where one or two atoms respectively are missing from the *closo* octahedron. It turns out that whereas *closo* species with n cage atoms require $(n+1)$ skeletal pairs as we have already described, their *nido* and *arachno* analogs with $n-1$ and $n-2$ cage atoms require $n+1$ skeletal bonding pairs too. Or, put another way, whereas *closo* species with n cage atoms require $(n+1)$ skeletal pairs, *nido* and *arachno* species with n cage atoms need $n+2$ and $n+3$ skeletal bonding pairs. It is as if the empty site behaves electronically as if it were there. Examples are given in Structure **11.16** and show some molecules from a different perspective. The series is isoelectronic with ferrocene. $C_6Me_6^+$ is simply a *nido* pentagonal bipyramid (number of skeletal pairs $= (1/2)[(6 \times 3) - 2] = 8$) with an apical CMe^{2+} unit; ferrocene is also a *nido* pentagonal bi-pyramid but contains the isolobal FeCp unit in place of CMe^{2+}. Isoelectronic $Mn(CO)_3$ may be used (Structure **11.6**) in its place to give $CpMn(CO)_3$. Structure **11.17** shows another series. Notice that C_4H_4 with six skeletal pairs is predicted to be a *nido* trigonal bipyramid (the tetrahedron) rather than the (Jahn–Teller unstable)

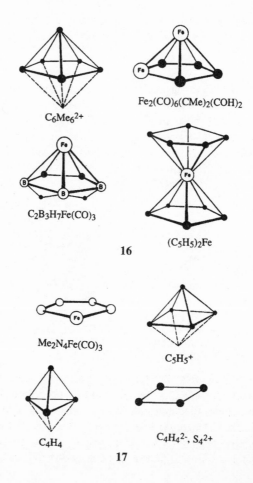

C₆Me₆²⁺

Fe₂(CO)₆(CMe)₂(COH)₂

C₂B₃H₇Fe(CO)₃

(C₅H₅)₂Fe

16

Me₂N₄Fe(CO)₃

C₅H₅⁺

C₄H₄

C₄H₄²⁻, S₄²⁺

17

benzvalene
18

square, which is the structure found (*arachno* octahedron) for $C_4H_4^{2-}$ or Se_4^{2+} (seven skeletal pairs). One intriguing example is shown in Structure **11.18**. This is the structure of benzvalene, C_6H_6, one of the known isomers of benzene. With $(1/2)(6 \times 3) = 9$ skeletal pairs it is an *arachno* dodecahedron.

Many of the clusters of both the transition metals and the boranes contain bridging hydrogen atoms, or small atoms inserted into the middle of the cluster. How do they influence the electron count?

The rôle of bridging hydrogen atoms in both types of cluster and central atoms in metal-based clusters such as Structure **11.2** and **11.5** are interesting. The additional atoms alter the counting rules of course since there are now extra orbitals to be included in the picture. An orbital diagram, which shows the effect of insertion of a carbon atom in an octahedral cluster is shown in Figure 11.2. (An analogous picture for the hydrogen-like cluster was

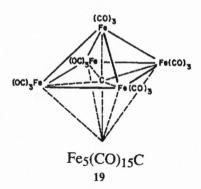

$Fe_5(CO)_{15}C$
19

shown in Figure 8.5.) Notice how the four carbon valence orbitals find symmetry matches with the cage orbitals so that there are still the same number of deep-lying orbitals. So the electron counting rules are preserved as long as the number of electrons from the inserted or bridging atoms is included in the count.

The interstitial or bridging atoms don't actually donate electrons to the cluster do they? This is what is implied by the way you count.

Just as in many of these counting rules the actual movement of electron density has little correlation with this formal counting procedure. In fact with a central carbon atom I suspect that electronegativity arguments would suggest charge transfer in the opposite direction. Clearly what has happened has been that metal–metal bonding within the cluster has been reduced but metal–carbon bonding increased on insertion of the central atom. An example of a molecule of this type is the *nido*-octahedral species $Fe_5(CO)_{15}C$ (Structure

11.19). Counting electrons in the way we have just described leads to: four electrons from the central carbon atom, 10 electrons from five $Fe(CO)_3$ units (two from each) leading to a total of 14 electrons (or seven pairs) for skeletal bonding. This is just the right number for an octahedrally based molecule. With five heavy atoms the molecule is a *nido* octahedron.

If there are bridging hydrogen atoms in the molecule then by analogy with our discussion around Figure 11.2, presumably we can regard them theoretically as each contributing one electron to the heavy atom core.

Yes, each bridging hydrogen atom contributes one electron to the skeletal electron count. The general counting rule noted above for a the number of skeletal electron pairs then becomes for the carborane $[(BH)_a(CH)_bH_c]^{-d}$ containing c hydrogen atoms $(1/2)(2a + 3b + c + d)$. So B_4H_{10} with seven skeletal pairs is an *arachno* octahedron. $Ru_6(CO)_{18}H_2$ is similarly a *closo* octahedron. Extra CO ligands behave analogously. With two electrons in its valence s orbital each additional CO molecule contributes two electrons to the skeletal count.

Are there exceptions to the rules?

Yes, there are. Obviously for the transition-metal containing fragments, the ligands need to be those with high Δ values so that the eighteen- electron rule applies. If this is not the

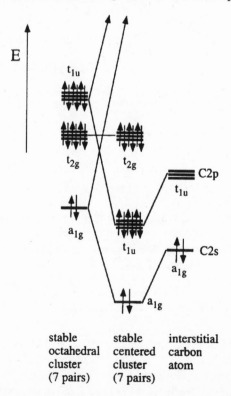

Figure 11.2
An orbital diagram showing how the valence s and p orbitals of an interstitial atom always find symmetry matches with the skeletal bonding orbitals of the deltahedron

case then other structures with different numbers of electrons are stable. There are many examples from oxide chemistry[6]. Even given this qualification concerning high-field ligands, there are exceptions to Wade's rules for systems that should perhaps at first sight obey them. Ni_6Cp_6 has a total of $(1/2)(6 \times 3) = 9$ skeletal pairs but is found with an octahedral nickel skeleton, rather than as an *arachno* dodecahedron. $(PhC)_4(BF)_2$, isoelectronic with $C_6Me_6^{2+}$, has a planar six-membered ring structure rather than the pentagonal pyramid found for the all-carbon analog. Its structure (see Structure **11.20**) may be stabilized by strong C=C linkages. In a similar fashion $C_4H_4^{2-}$, obeys Wade's rules and is found as the square planar *arachno* octahedron. This structure may be similarly favored over the cis-divacant octahedron found for B_4H_{10} by C–C π interactions. By and large however, the rules work well. Of course there has been much activity to make molecules that violate the rules and in understanding why they do.

20

References

1. Wade, K., *Chem. Commun.*, 792 (1971); *Inorg. Nucl. Chem. Lett.*, **8**, 559, 563 (1972); *Adv. Inorg. Chem. Radiochem.*, **18**, 1 (1976).
2. Mingos, D. M. P., *Nature (Phys. Sci.)*, **236**, 99 (1972); *Chem. Soc. Rev.*, **15**, 31 (1986).
3. Stone, A. J., *Inorg. Chem.*, **20**, 563 (1981); Stone, A. J. and Alderton, M. J., *Inorg. Chem.*, **21**, 2297 (1982).
4. Mingos, D. M. P. and Wales, D. J., *Introduction to Cluster Chemistry*, Prentice-Hall (1990).
5. Longuet-Higgins, H. C. and Roberts, M. de V., *Proc. Roy. Soc.*, **224**, 336 (1954).
6. McCarley, R. E., *Inorganic Chemistry: Toward the 21st Century*, Chisholm, M. H. (ed), ACS Symposium Series, No. 211, American Chemical Society (1983).
7. Cotton, F. A. and Wilkinson, G., *Inorganic Chemistry*, Fifth Edition, John Wiley & Sons (1988).

General References and Further Reading

Johnson, B. F. G., *Transition Metal Clusters*, John Wiley & Sons (1979).
Rudolph, R. W., *Accts. Chem. Res.*, **9**, 446 (1976).
Cotton, F. A. and Wilkinson, G., *Inorganic Chemistry*, fifth Edition, John Wiley & Sons (1988).

12 Is Delocalization of Electrons always Stabilizing? Or, Why is Benzene a Regular Hexagon?

One of the fundamental ideas, certainly of the organic chemical world, but also in other areas of chemistry is that delocalization of electrons is energetically beneficial for molecules. It provides a way of stabilizing, by resonance, symmetrical structures such as those of benzene (Structure **12.1**), and the allyl cation (Structure **12.2**) through their π systems. The ideas come from valence bond theory and are well rooted in our science. There are similar situations in inorganic chemistry, the linear tri-iodide anion being an example. But if such delocalization is so energetically important, why is there no stable hexaazabenzene (N_6) molecule, isoelectronic with benzene itself or no stable cyclic H_6 or linear H_3^- molecules, where such resonance stabilization is also possible? What are the electronic ingredients that lead to symmetrical structures for some molecules containing delocalized electrons (benzene or tri-iodide ion and the alkali metal trimers) but the absence of such arrangements for many isoelectronic analogs. Is the simple picture most of us have concerning resonance stabilization too simple?

This is a question that has lain at the heart of chemistry for many years, indeed from the earliest days of the concept of aromaticity[1]. Just why does benzene have a regular hexagonal structure? In more modern language, and more generally, just how important is resonance stabilization in determining the observed structures of molecules such as these? The idea of resonance stabilization shown in Structure **12.3** for benzene (and mentioned in Chapter 3) is a simple one to appreciate. The interaction of the valence bond wavefunctions describing the two localized canonical (Kekulé) forms (Structure **12.4**) lead to an energetic stabilization (the quantum mechanical resonance energy, QMRE). There are other canonical forms that should be included too, but these two will do for the sake of the argument here. The result is delocalization of electrons around the benzene ring or over all three carbon atoms in allyl cation. The organic chemist writes the structures of benzene and the allyl cation as in Structure **12.5** as a result. For many years now this has been considered[2] as being the energetic effect that sets the symmetrical structure of benzene, as a perusal through many organic chemistry texts (see, for example, reference 3) will show. As you point out, although these molecules are unknown, a similar resonance stabilization can be envisaged for isoelectronic cyclic N_6 and for cyclic H_6. For the latter, the σ system plays the same rôle as the π levels in benzene and N_6. In a similar vein, the

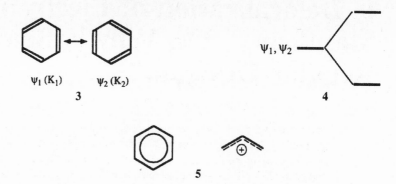

$\psi_1 (K_1)$ $\psi_2 (K_2)$

3

4

5

molecules H_3 and H_3^- do not exist, whereas symmetric clusters of alkali metals, the trihalide ions and the symmetric allyl cation geometry are known.

The true story is actually quite different from this traditional one. In fact we will see that for many species electronic delocalization is forced, with the resonating π system trapped as an unstable transition state in a stiff σ framework. The discussion here follows work of Phillippe Hiberty, Sason Shaik and their co-workers[4].

The first thing to note is that the internuclear separations in the delocalized systems are larger than in the localized ones (e.g., I_2 $r(I-I) = 2.67$ Å, I_3^- $r(I-I) = 2.91$ Å; typical C=C distance = 1.33 Å, benzene $r(C-C) = 1.39$ Å). Thus an important first step will be to see what happens energetically on changing internuclear distances. Figure 12.1 shows schematically how the energies of the two benzene Kekulé forms (Structure **12.3**) labeled as K_1 and K_2, change as the π double bonds are stretched. Although our discussion will apply to benzene, analogous arguments apply for the allyl cation as in Structure **12.6** or the tri-iodide anion. At the point R where the internuclear distances are equal, the regular structure is found (this is the regular hexagon for benzene, the symmetrical structure for allyl cation or tri-iodide anion). Notice that both K_1 and K_2 are destabilized during this bond stretching process at the point R. All canonical structures will be destabilized in this way. Figure 12.1 also shows how the two Kekulé structures interact with each other as in Structure **12.4** leading to a resonance stabilization that is maximized on energy gap arguments at the point R.

6

So the model suggests two possibilities (Figure 12.2). If the quantum mechanical resonance energy (B in Figure 12.1) is sufficient to overcome the energetic penalty associated with the bond stretching, then, as shown in Figure 12.2(a), a stable resonating state may occur at the symmetrical point R. If the converse is true, then as shown in Figure 12.2(b), the resonating state is unstable. But what determines the magnitudes of the terms involved?

An important player is the energy separation G (Figure 12.1). The smaller G, the larger one expects the interaction to be between the two states K_1 and K_2. G depends on ΔE_T, the singlet–triplet energy difference in X_2, the lowest energy electronic excitation (related to the HOMO– LUMO gap), in turn related to the contribution to the X_2 bond energy ($D(X_2)$)

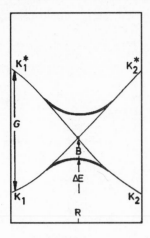

Figure 12.1
A schematic that shows how the two benzene Kekulé forms (K_1 and K_2) change in energy as the bonds are stretched. Both K_1 and K_2 are destabilized during this process but interact with each other via resonance, maximized on energy gap arguments at the point R

by the electrons under consideration. Another intuitively useful parameter is the ratio $\Delta r / r(X_2)$. Δr is the extent of bond stretching, needed to get from K_1 or K_2 to R, and $r(X_2)$ the equilibrium distance in the Kekulé structure. This is a measure of the cost in energy associated with the distortion on moving to the point R from K_1 or K_2. Table 12.1 shows the results of some calculations[4]. Only one orbital manifold is involved for the first three systems, the set of 1s orbitals for hydrogen. For lithium the calculations are broken down into two parts, one for the set of 2s orbitals and another for the $2p\pi$ orbitals. The calculation on the Li 2p system represents that for a suspended π system.

The destabilization of the symmetric structure scales in the way expected with all three parameters for the first three systems. Thus a general prediction might be that delocalized X_n systems should only be stable when the X_2 molecule is weakly bound itself. In these cases there is only a small energetic penalty on stretching the X—X bond.

So we should not expect to find symmetric H_6 molecules coordinated to transition metal fragments in the way one finds coordinated benzene or cyclopentadienyl molecules.

Yes, none have yet been made. It is certainly a way too of understanding the non-existence of linear H_3 and H_3^-. The bond energy in H_2 is very high as you imply in your

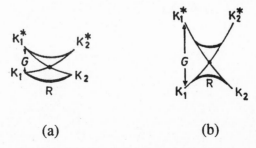

(a)   (b)

Figure 12.2
Two possibilities that arise from the picture of Figure 12.1: (a) a stable resonating state at the symmetrical point R will be found if the quantum mechanical resonance energy (B) is larger than the energetic penalty associated with the bond stretching; (b) if the converse is true the resonating state is unstable

Table 12.1 Calculated properties of delocalized systems (CI level[a]) (energies in kJ/mol[b])

X_6	$\Delta E(3X_2 \rightarrow X_6)$	$\Delta r/r(X_2)$	ΔE_T	$D(X_2)$
H_6	$+455$	34.5	1610	535[c]
$Li_6(\pi)$	$+115$	26.6	200	55
$Li_6(\sigma)$	$+0.2$	7.0	105	45
N_6	$\sim 0(\sigma+\pi)$	7.6	410[d]	940
C_6H_6	$<0(\sigma+\pi)$	4.8	450[d]	305

[a] See Chapter 5.
[b] From ref. 4 (energies given to nearest 5 kJ/mol).
[c] Note that these are calculated values. The experimental H–H bond energy is 435 kJ/mol.
[d] These are triplet excitation energies in $HN=NH$ and $H_2C=CH_2$ respectively.

question. However, I_3^- is known, a fact not unconnected with the low bond dissociation energy of I_2. Recall however, that H_3^+ is known as a regular triangular molecule, 'isoelectronic' with $C_3H_3^+$. (As an aside, perhaps one could make a molecule, such as $V(CO)_5(H_3)$, $(V(CO)_5^-(H_3^+))$ with an eighteen-electron count at the metal that would be stable.)

The calculations for the last two systems included both σ and π orbital manifolds. The calculated sign of the dissociation energy, $\Delta E(3X_2 \rightarrow X_6)$ is in accord with observation for benzene. That for N_6 is close to zero and certainly indicates that the molecule is either not stable or is only marginally so.

However, the magnitude of ΔE_T for the π systems of these two molecules, by comparison with values for the first three examples, suggests that the regular structure should certainly be unstable here too. We know though that the structure of benzene is a regular hexagon.

The answer to this problem is to be found in the results shown in Table 12.2[4] that gives a breakdown of $\Delta E(3X_2 \rightarrow X_6)$ into σ and π contributions for some four- and six-membered rings. Notice that the sign (negative) of $\Delta E(\sigma)$ shows that the σ manifold always favors the regular structure, but that of $\Delta E(\pi)$ shows that the π system always disfavors it. So, although the QMRE associated with the π manifold is always positive, i.e., resonance is always stabilizing, the data show that it is not large enough to overcome the energetic penalty associated with the stretching of bonds required to reach the regular geometry.

Table 12.2 Calculated distortion energies in four- and six-membered rings (CI level) (energies in kJ/mol[a])

X_6	$\Delta E(\sigma)$	$\Delta E(\pi)$	$\Delta E(\text{total})$	QMRE[c]	ΔE_T
N_6	-55	$+55^b$	0^b	430	410
C_6H_6	-70	$+40$	-30	355	450
P_6	-15	$+10$	-5	185	155[d]
Si_6H_6	-20	$+10$	-10	175	170[d]
C_4H_4	-30	$+45$	$+15$	125	415
Si_4H_4	-10	$+10$	0	75	170

[a] From reference 4 (energies given to nearest 5 kJ/mol).
[b] $\Delta E(\pi) = 60$ kJ/mol from an SCF calculation without CI. Thus the sign of $\Delta E(\text{total})$ depends on the nature of the calculation.
[c] The Quantum Mechanical Resonance Energy B (Figure 12.1).
[d] The actual values are basis-set dependent.

It's very interesting to see that the symmetric structure of benzene arises through the energetic preferences of the σ framework rather than via π resonance stabilization.

Yes. It's a conclusion somewhat different to that usually encountered. Notice that for cyclobutadiene the energetic demands of a (pseudo) Jahn–Teller unstable π manifold (see Chapter 14 for a discussion of the various types of Jahn–Teller effects) outweighs the preference of the σ manifold for the symmetric structure. In the second row analog Si_4H_4 where π effects are much less important, this is not true and the square is predicted to be the stable structure. (Parenthetically we note that hexagonal benzene is second-order Jahn–Teller unstable via the π network, a weaker energetic effect.)

A useful model to view the preferences of the σ manifold in general is to describe the interatomic interactions by a harmonic potential, $V = 1/2kx^2$, where x is the interatomic separation. If alternate X–X linkages are respectively stretched and contracted (by Δx) then the energy is $V = 1/2k(\Delta x)^2 + 1/2k(-\Delta x)^2$, minimized at the symmetric structure where $\Delta x = 0$ and where all bond lengths are equal. We know that a σ skeleton behaves this way from experience. Recall the (approximately) equal C–C distances in saturated hydrocarbons.

It all sounds very convincing, but these results surely depend on the way we distribute the various electronic contributions to the energy.

This is quite true. The approach we have described, in fact initially used a method where one has to make a choice as to how to divide the electronic repulsion between the σ and π electrons between the two manifolds. The assumption made above was to put it entirely into the σ framework. Obviously there are various ways to do this. The result of several studies however, show that the general conclusions are correct, although the numerical details may differ. The most recent results[4] show that the π manifold of benzene is indeed unstable to distortion in exactly the same way as the σ framework of cyclic H_6. It does bring up, though, an important aspect of many theoretical treatments that is relevant for many of the problems in this book. Questions of this type do not have *rigorous* answers since they do not involve a physical observable, i.e., one that corresponds to a well-defined quantum mechanical operator. Many concepts in chemistry fall into this category, electronegativity for example. But they are, these qualifications not withstanding, very useful tools with which to understand a wide variety of chemical phenomena.

I suppose we could force the C–C distances in benzene to be inequivalent just by applying strain to the molecule.

Yes, Structure **12.7** shows one[5] of the molecules that have been used to do this. The ring is flat, the CCC angles are 120° and the C–C distances alternate between 1.44 Å and 1.35 Å. But results such as these make little comment about what is really going on.

Now presumably this electronic model can be used in other areas of chemistry.

Yes, of course, just like any useful model. One particularly nice example shows[6] why there are many examples of trigonal bi-pyramidal SiX_5^- species as stable compounds, while analogous CX_5^- species are quite rare as stable species. (The geometry occurs, of course, as a transition state in the S_N2 reaction.) The calculations are easiest to do for

7

CH_5^- and SiH_5^- and the results are shown in a simplified form in Figures 12.3 and 12.4. The form of the abscissa is shown in Structure **12.8**. There is a larger computed resonance stabilization energy for silicon compared to carbon, and a lower energetic penalty paid by the silicon compound on distortion leading to different behavior in the two cases. Weaker Si—H compared to C—H linkages is one contribution to this result. (Electronically the

$$Y + \quad \overset{\diagdown}{\underset{\diagup}{A}}\!\!-\!\!X \rightarrow Y\!\!-\!\!\overset{|}{\underset{|}{A}}\!\!-\!\!X \rightarrow Y\!\!-\!\!A\overset{\diagup}{\diagdown} \quad + X$$

point R

a   b   c

8

picture is more complex[6] than that of Figure 12.1, but the basic results easy to understand.)

We can also use this result to gain insight into the paucity of compounds containing 'five-valent' carbon in general (see Chapter 6). It is interesting to see it is a model that does not use the absence of valence d orbitals on carbon as a focal point.

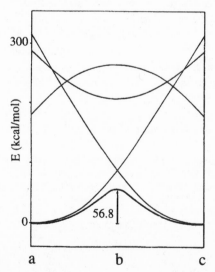

Figure 12.3
A calculated plot of the type shown in Figure 12.1 for CH_5^-. The points **a**, **b**, **c** *correspond to those in Structure* **12.8**

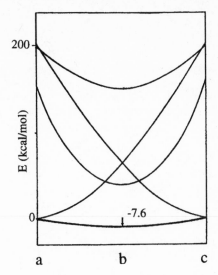

Figure 12.4
A calculated plot of the type shown in Figure 12.1 for SiH$_5^-$. The points **a**, **b**, **c** *correspond to those in Structure* **12.8**

References

1. Kekulé, A., *Bull. Soc. Chim. Fr.*, **3**, 98 (1865).
2. Pauling, L., *The Nature of the Chemical Bond*, Third Edition, Cornell University Press (1960).
3. E.g. Morrison, R. T. and Boyd R. N., *Organic Chemistry*, Sixth Edition, Prentice Hall (1992), p 500.
4. Shaik, S. S. and Hiberty, P. C., *J. Amer. Chem. Soc.*, **107**, 3089 (1985); Ohanessian, G., Hiberty, P. C., Lefour, J.-M., Flament, J.-P. and Shaik, S. S., *Inorg. Chem.*, **27**, 2219 (1988); Hiberty, P. C., Danovich, D, Shurki, A. and Shaik, S. S., *J. Amer. Chem. Soc.*, **117**, 7760 (1995); Shaik, S. S. and Bar, R., *Nouv. J. Chim.*, **8**, 411 (1984).
5. Frank. N. L., Baldridge, K. K. and Siegal, J. S., *J. Amer. Chem. Soc.*, **117**, 2102 (1995).
6. Sini, G., Ohanessian, G., Hiberty, P. C. and Shaik, S. S., *J. Amer. Chem. Soc.*, **112**, 1407 (1990).

13   What is the Origin of Hückel's Rule?

The relative stability of cyclic polyenes as a function of electron count, is a question of longstanding interest. Erich Hückel in 1936 proposed a rule (the $4n+2$ rule) based on the orbital structure of rings of carbon pπ orbitals that he calculated using the molecular orbital approach that bears his name. Using such simple Hückel theory, the energy levels of cyclic systems take on a particularly simple form. For a cyclic system with N atoms, the energy of the jth orbital is given by

$$E_j = \alpha + 2\beta \cos 2j\pi/N \qquad (13.1)$$

where j runs from $0, \pm1, \pm2 \ldots (\pm N/2$ for N even) or $((N-1)/2$ for N odd). Notice in general that this means that there is always a single level with an energy of $\alpha + 2\beta$ at lowest energy and that all the other energy levels occur in pairs, except that the very highest energy level for even-membered rings occurs as a singleton at $\alpha - 2\beta$ (Structure **2.5**). Thus Hückel's rule just reflects the generation of closed shells of bonding or bonding plus non-bonding levels (Structure **13.1**). Is there more to it than this?

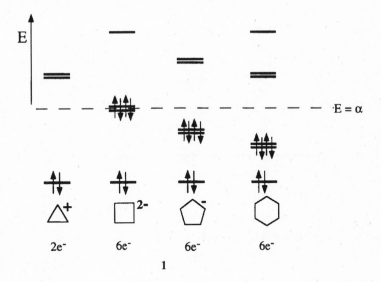

1

Hückel's rule is of course determined by quantum mechanics[1] but there is a significant input from topological considerations here too, namely it is how the atoms are linked together which determines the form of the secular determinant and hence the location of the energy levels. Hückel theory is therefore often called a 'topological' theory, a point you made in Chapter 2. Figure 13.1 shows one aspect of this. First label each of the atoms of the π-network into two types, starred and unstarred, and try to arrange the stars such that no two starred, or no two unstarred atoms lie next to each other. If you can do this (as in Structures **13.2** and **13.3**) then we call the molecule an alternant hydrocarbon. If such pairs

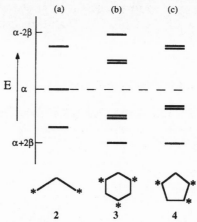

Figure 13.1
Hückel molecular orbital diagrams for (a) odd; and (b) even alternant hydrocarbons; (c) non-alternant hydrocarbon, 13.2–13.4 respectively

of atoms have to lie together because of the way the atoms are linked together, then a non-alternant hydrocarbon results (Structure **13.4**). Alternant hydrocarbons come in two types, those where the numbers of starred and unstarred atoms are equal (an even alternant hydrocarbon, Structure **13.3**) and those where they are not (an odd alternant hydrocarbon, Structure **13.2**). Figure 13.1 shows Hückel diagrams for the $p\pi$ levels of these three systems. Even alternant hydrocarbons possess a set of levels symmetrically located about $E = \alpha$. For the odd-alternant hydrocarbon there will be non-bonding energy levels located at $E = \alpha$. There is one in the example shown. For the non-alternant system there is no mirror symmetry of the molecular orbital pattern about this point. A three-dimensional Hückel theory is behind the structures of the cage and cluster molecules of Chapter 11. These topological considerations have attracted quite a bit of attention from graph theorists, specifically interested in properties set by atomic connectivity rather than the detailed form of the electronic energy levels. In fact many of the comments made about electronic–structural problems in this book reflect the extremely important rôle played by symmetry and topology.

Are there any insightful ways to investigate this 'topological' nature of the theory?

One particularly useful method that highlights this connectivity approach comes from the concept of the moments[2,3] of the electronic density of states. We used the second moment in our discussions in Chapter 5. The nth moment of the collection of energy levels $\{E_i\}$ is given by the expression

$$\mu_n = \sum_i E_i^{\,n} \tag{13.2}$$

A very useful way of writing the moments is in terms of the H_{ij} integrals or Hückel β's of the orbital problem. The nth moment of the electronic density of states is just the trace, Tr, or the sum, of the diagonal elements, of H_n where H is the Hamiltonian matrix of Chapter (5). There is a straightforward way to see this. Since H is Hermitian there is a unitary matrix S such that $S^{-1}HS$ is diagonal and whose diagonal elements are the eigenvalues, or orbital energies, of H. Since $\text{Tr}(H_n) = \text{Tr}[(S^{-1}HS)]$ and $S^{-1}HS$ is diagonal with elements E_i, then $\sum_i E_i^{\,n} = \text{Tr}(H_n)$. We can write $\text{Tr}(H_n)$ in terms of the H_{ij} integrals to get a very interesting geometrical view of μ_n. As shown in equation (13.3) the nth moment is just a weighted sum over all the self-returning walks of the orbital problem:

$$\mu_n = \sum_{\text{walks of length } n} H_{ij}H_{jk}\ldots H_{ni} \tag{13.3}$$

So this 'topological' result, shown pictorially in Structure **13.5**, reveals a very useful connection between a function of the electronic energy levels, the μ_n, and the geometry via the orbital connectivity.

5 $2\beta^2$ $6\beta^2$ **6**

Yes. There are only contributions to μ_n from those walks that connect atoms with non-zero values of H_{ij}, namely those linked together by the σ framework using the Hückel model. The first moment is the sum of the 'walks-in-place' and is thus set by the values of the H_{ii} or α terms. To simplify our discussion we can easily put it equal to zero since it just sets the 'zero' of the energy scale. Using this simplification the second moment (equation (13.4)) is just the sum of the squares of the interaction integrals that tie one orbital (i) to its neighbors (j) as shown in Structure **13.6**:

$$\mu_2 = \sum H_{ij}H_{ji} = \sum H_{ij}^2 \tag{13.4}$$

Thus one view of the second moment is a measure of the coordination strength around an atom.

So higher moments must describe in a similar way how a given atom electronically 'sees' its neighbors further away.

Yes, these are shown in Structures **13.7** and **13.8**. Although we do show how the third and fourth moments give information, via returning walks through the orbital structure,

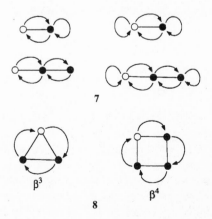

7

β^3 β^4

8

124 WHAT IS THE ORIGIN OF HÜCKEL'S RULE?

about the electronegativity difference between the atoms at each end of the bond, it is their contributions from rings of various sizes that is of particular importance here. Thus the four-ring contains contributions to the fourth moment via returning walks around the ring of length 4 (Structure **13.8**) which are not found in open four-atom systems. We expect that the influence of an atom on its neighbors will decrease as the distance between them increases, so that the importance of the higher moments should decrease as their order increases.

We are now in a position to start to examine the origin of the energy differences between cyclic π systems of different sizes[2,3] as a function of electron count (Hückel's rule). Figure 13.2 shows the computed energy differences between two six-rings and p 12/p rings as a function of electron count using the Hückel model. (We can envisage two parameters for the abscissa. One is just the number of pπ electrons, which we show, the other a more general one, the fractional orbital occupancy x, which runs from 0 to 1.)

> I expect you chose this comparison since it is important (Chapter 5) to work at constant second moment and thus all of the energy differences are between systems with the same local coordination number.

Exactly. Notice two important features. The first is that the amplitudes of the two curves decrease in the order, three rings > four rings. We can understand this result in the light of the discussion above. The energetic weights of the closest neighbors are most important. Clearly, you can see that when comparing a p-membered ring with the 6-membered ring ($p < 6$) the first self-returning walk (moment) which will be different between the two will be the walk that goes around the p-membered ring of length p back to the starting atom (as shown for the three-, four-membered rings in Structure **13.8**, for example). Using this moments language the two plots of Figure 13.2, labeled by triangle and square, represent third- and fourth-moment problems respectively. Another important result is that the number of nodes in these curves (including those at $x = 0$, 1) is equal to the order of this first disparate moment between the two structures. It is the structure with the largest moment (largest number of walks) which is the one that is more stable at the earliest orbital occupancy.

We have therefore shown some very interesting topological results. The shape and amplitude of these energy difference curves between two structures, $\Delta E(x)$, as a function of fractional orbital filling, x, are determined quite simply by the way the atoms are connected

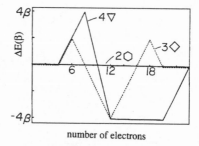

Figure 13.2
Computed energy differences between two six-rings and p 12/p rings as a function of electron count using the Hückel model. (Two parameters can be used for the abscissa, the number of pπ electrons (shown), and the fractional orbital occupancy x, which runs from 0 to 1

together. It is this atomic connectivity that determines the orbital walks that are possible and hence the form of the energy level pattern as measured by the set of moments.

So we can use the returning walks around the ring to comment on the stability of rings of different sizes.

Yes, let's see what Figure 13.2 tells us about Hückel's rule. Remember that we are comparing the energies of two six-membered rings with p $12/p$-membered rings. Three-membered rings are most stable at early electron counts with a maximum in their stability curve at eight electrons ($x = 0.33$) or $8/4 = 2$ electrons per three-membered ring. This is just the point where there is the Hückel's rule quota of $(2n + 1)$ electron pairs per π-system. ($n = 0$ in this case). There are two regions of stability for the four-membered rings, at $x = 0.25, 0.75$. These correspond to totals of 6 and 18 electrons or two and six electrons per four-membered ring. These are just the electron counts expected from Hückel's rule with $n = 0, 1$. Although the calculations show the stability of rings compared to the open-chain compound (approximated here by the two six-membered rings) the results have much more important consequences. They tell us in very general electronic terms when particular structural features should be stable. At low electron counts (and especially at one-third filling), the three-rings are stable, and particularly so given the amplitude of the plot at this point. At the half-filled point, six-membered rings are stable, and at higher electron counts (especially around three-quarters filling) four-rings are stable. A set of $\Delta E(x)$ curves[2], obtained by treating this problem much more generally, is shown in Figure 13.3. Here you can see clearly for the six-membered rings the three regions of stability at $x = 0.167, 0.5$ and 0.833, or two, six and ten electrons per ring appropriate to the Hückel values of 0, 1, 2.

These ideas must be far broader in scope than Hückel's rule. The arguments you have given are completely general. So although you have used carbon $p\pi$ orbitals in the cyclic hydrocarbons as an example, there is no reason why other orbitals can't be used in the moments description in other parts of chemistry.

Indeed. We know that many cage and cluster boranes and metal clusters, often (but incorrectly) regarded as electron deficient have structures that are built up from three rings (Chapter 11). This is a natural consequence of these results too. They have values of x that

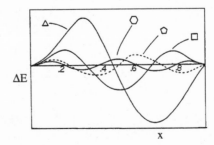

Figure 13.3
Typical $\Delta E(x)$ curves as a function of fractional orbital (molecule) or band (solid) filling. (One of them is dashed for clarity.) These are the curves expected for the energy difference between two structures whose first disparate moment in the energy density of states is the mth (m = 3–6)

are much less than 0.5. They are 'deficient' after all. Similarly, notice the shape of the energy difference curves for close-packed structures (of atoms but of three-rings too) of elemental Li and Be in Figure 5.6. They have more three-rings than does rhombohedral boron R-12, the reference structure. As we move past three electrons per atom toward the half-filled point, the diamond and graphite structures with their six-membered rings become the most stable structures.

An interesting series of structures that underline your comment concerning the versatility of the ideas is that of the heavy elements at the right-hand side of the Periodic

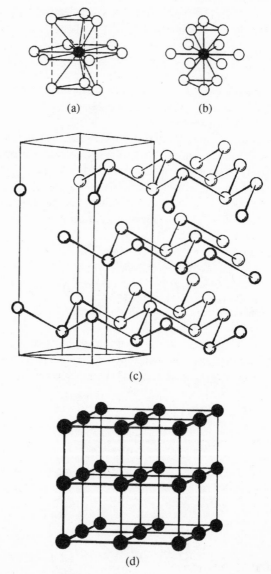

(a)   (b)

(c)

(d)

Figure 13.4

The structures of the elements Tl–Po: (a) hexagonal closest packing (Tl); (b) cubic closest packing (Pb); (c) α-arsenic (Bi), (d) simple cubic (Po)

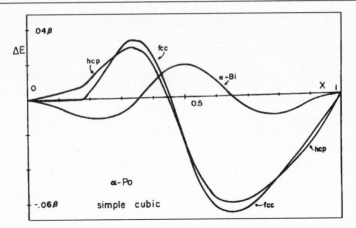

Figure 13.5
Energy difference curves relative to that of the simple cubic structure (Po), for the elemental structures Tl–Po

Table, Tl–Po. As Figure 13.4 shows, Tl and Pb have close-packed structures where three-rings are in abundance. In fact these structures could be regarded as the closest packing of three-membered rings as we have just noted. Bi has the α-arsenic structure that contains no ring smaller than six and Po has the simple cubic structure that contains four- rings. Figure 13.5 shows the results of some calculations designed to investigate the electron count dependence on structure. Since at the bottom of the Periodic Table the $6s^2$ pair of electrons are often regarded as inert, the calculations employed the 6p orbitals only. Notice how the stable structure is reproduced by calculation in each case and also that the form of the energy difference curves is completely in keeping with the size of the smallest rings in the structure.

There is an important message here. There is much to be gained in looking at the energy difference curves between two structures over the whole range of electron occupancy, rather than focusing on the energy difference between two structures at a single electron count. Also, study of extended features in the structure, especially three-, four- and six-membered rings, is important too. One should not focus entirely on the local coordination.

Yes, but it's not always this simple to analyze. If several important orbital manifolds overlap in energy (such as π and σ), then the $\Delta E(x)$ curves are not always so simple to sort out.

References

1. Streitweiser, A., *Molecular Orbital Theory for Organic Chemists*, John Wiley & Sons (1961).
2. Lee, S., *Accts. Chem Res.*, **24**, 249 (1991).
3. Burdett, J. K., *Structure and Bonding*, **65**, 30 (1987).

General References and Further Reading

Coulson, C. A. and Longuet-Higgins, H. C., *Proc. Roy. Soc.*, **A191**, 39 (1947), **A192**, 16 (1947), **193**, 447, 456 (1948).

Heilbronner, E. and Bock, H., *The HMO Model and its Applications*, John Wiley & Sons (1976).

Hückel, E., *Z. Phys.*, **70**, 204 (1931).

McGlynn, S. P., Vanquickenborne, L. G., Kinoshita, M. and Carroll, G. G., *Introduction to Applied Quantum Chemistry*, Holt, Rheinhart and Winston (1972).

Streitweiser, A., *Molecular Orbital Theory for Organic Chemists*, John Wiley & Sons (1961).

Yates, K., *Hückel Molecular Orbital Theory*, Academic Press (1978).

14 Is the Jahn–Teller Effect as Simple as it Seems?

The Jahn–Teller theorem tells us that degenerate electronic states in non-linear molecules are unstable with respect to a geometrical distortion that lifts the degeneracy. The classic inorganic case of a Jahn–Teller molecule is that of octahedral Cu(II). Thus the copper atom in $CuBr_2$ lies in a distorted environment (Structure **14.1**). However, the copper atom also lies in a distorted environment in α-$CuBr_2(NH_3)_2$ (Structure **14.2**), but here, since the local symmetry is far from O_h, there can never be a degenerate electronic state (except by accident) for the structure with all bond lengths at their 'normal' values.

Cyclobutadiene is the example of a Jahn–Teller unstable molecule that comes to mind from organic chemistry. It has two electrons in a doubly degenerate orbital, but when I evaluate the symmetry species of the electronic states which this configuration demands, none of them are degenerate. It appears that what we call the Jahn–Teller Theorem is somewhat more complex than appears at first sight.

It is certainly true that many of the structural distortions labeled as 'Jahn–Teller' distortions actually have quite different theoretical arguments behind them than those initially envisaged by Jahn and Teller[1] although in many cases they are connected. Figure 14.1 shows the behavior expected. On distortion (along a coordinate q) the degenerate electronic state splits apart in energy and the lower energy branch has an energy minimum away from the symmetric geometry ($q = 0$). Always be careful though to remember that these are symmetry-based results. They make no predictions concerning energetics. One should always be careful too, not to ascribe Jahn–Teller distortions to all distorted molecules. For example[2] the series of molecules $MX_3(NMe_3)_2$ (M = Ti, V, Cr; X = halide) are all based on a trigonal bipyramidal geometry and all are distorted. The magnitude of the distortion increases in the order V < Ti < Cr. One way to identify Jahn–Teller unstable systems has traditionally been to test for asymmetric occupation of degenerate orbitals. So, from Structure **14.3** we can see that the electronic state of the vanadium complex is certainly not a candidate for distortion, but the other two are. The larger distortion for Cr compared to Ti may be understood as arising from an orbital degeneracy involving the σ manifold for M = Cr but one involving the electronically less potent π manifold for Ti. The smallest distortion for M = V probably arises through forces normally described as 'crystal packing effects'. These distortions just represent the result of packing together pliable molecules in the solid.

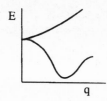

Figure 14.1

The behavior expected for a (first-order) Jahn–Teller distortion. Distortion along the coordinate q splits the degenerate electronic state present at q = 0 apart in energy. The lower energy branch has an energy minimum away from the symmetric geometry (q = 0)

We can use and extend the symmetry arguments of Jahn and Teller to put their ideas into perspective that will enable us to understand details of molecular geometry. First of all let's point out that there are in fact three different types of 'Jahn–Teller' distortion. Described as

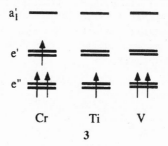

first-order, second-order and pseudo Jahn–Teller distortions, they are applicable to three distinctly different electronic situations. (The second and third are often both called pseudo Jahn–Teller effects, an unfortunate notational choice that blurs the differences in their origin.) The mathematical basis of all three comes from the perturbation expansion[2,3,4] of the electronic energy, $E_0(q)$, along a distortion coordinate q:

$$\mathcal{H}(q) = \mathcal{H}^0(0) + (\partial\mathcal{H}/\partial q)_0 q + (1/2)(\partial^2\mathcal{H}/\partial q^2)_0 q^2 + \dots$$
$$E_0(q) = E_0^0(0) + E_0^1(q) + E_0^2(q) + \dots \tag{14.1}$$

The perturbation is just $\mathcal{H}' = (\partial\mathcal{H}/\partial q)_0 q + (1/2)(\partial^2\mathcal{H}/\partial q^2)_0 q^2$ if we just keep terms up to those of order q^2. So the first-order correction to the energy is

$$E_0^1(q) = \langle 0|\mathcal{H}'|0\rangle = \langle 0|\partial\mathcal{H}/\partial q|0\rangle q + (1/2)[\langle 0|\partial^2\mathcal{H}/\partial q^2|0\rangle q^2 \tag{14.2}$$

and the second-order correction is

$$E_0^2(q) = \sum_j{}' |\langle 0|\partial\mathcal{H}/\partial q|j\rangle|^2/(E_j(0) - E_0(0))]q^2 + \dots \tag{14.3}$$

keeping terms no higher than q^2. Collecting terms in q and q^2 leads to equation (14.3) for the energy of the electronic ground state $|0\rangle$.

$$= E_0^0(0) + |\langle 0|\partial\mathcal{H}/\partial q|0\rangle|q + (1/2)[\langle 0|\partial^2\mathcal{H}/\partial q^2|0\rangle$$
$$- \sum_j{}' |\langle 0|\partial\mathcal{H}/\partial q|j\rangle|^2/(E_j(0) - E_0(0))]q^2 + \dots \tag{14.4}$$

Although this looks complex, the details of the integrals are not of much interest. Symmetry considerations are much more important in the implementation of this equation. $E_0^0(0)$ is an energy zero term that we can ignore since we are interested in energy changes away from the reference geometry ($q = 0$). $E_0^1(q) = \langle 0|\partial\mathscr{H}/\partial q|0\rangle$ is the term first-order in q.

As Jahn and Teller pointed out, for $\langle 0|\partial\mathscr{H}/\partial q|0\rangle$ to be non-zero, Γ_q, the symmetry species of q, must be contained within the symmetric direct product of the symmetry species of $|0\rangle$, Γ_0. (The Hamiltonian operator is totally symmetric.)

Yes, Jahn and Teller showed by evaluation for many point groups that this is only true for degenerate electronic states of non-linear molecules. Thus orbitally degenerate electronic states of non-linear molecules will distort to remove this degeneracy, the usual statement of the theorem. The phrase 'orbitally degenerate' is included to eliminate degeneracies arising via spin. The symmetry of $|0\rangle$, Γ_0, controls the identity of q, the 'Jahn–Teller active' mode, since once Γ_0 is set, Γ_q is automatically determined if $\langle 0|\partial\mathscr{H}/\partial q|0\rangle$ is to be non-zero. For an E_g state of an octahedral molecule (point group O_h), the symmetric direct product of Γ_0 (e_g) is $a_{1g} + e_g$. For $\langle 0|\partial\mathscr{H}/\partial q|0\rangle$ to be non-zero then $\Gamma_q = a_{1g} + e_g$ too. A motion of a_{1g} symmetry does not change the geometry in a way to remove the degeneracy (it's a totally symmetric stretching mode), and may thus be discarded, but one of e_g symmetry does. Thus the Jahn–Teller active mode is of e_g symmetry, $\langle 0|\partial\mathscr{H}/\partial q|0\rangle$ is non-zero by symmetry (the theory makes no comment on its magnitude) and there is a driving force way from the reference geometry with $q = 0$. This is then the first-order Jahn–Teller theorem. If $|0\rangle$ is non-degenerate (an A or B state) then Γ_q always only contains the totally symmetric representation ($a \otimes a = a_1$, $b \otimes b = a_1$) and thus the molecule is first-order Jahn–Teller stable.

Typical inorganic examples of first-order Jahn–Teller unstable systems are octahedral transition metal MX_6 molecules with low-spin d^7, d^8 or d^9 configurations. (Actually, I think the low-spin d^8 case is rarely viewed as a Jahn–Teller distortion, but it is.) The degenerate electronic states are ^2E_g (e_g^1), ^1E_g (e_g^2) and ^2E_g (e_g^3) respectively with the prediction therefore of a distortion coordinate of e_g symmetry in each case.

Yes, two components of a stretching motion of this symmetry are shown in Structure **14.4**. Thus we say that octahedral Cu(II) or low-spin Ni(II) complexes with a local MX_6 stoichiometry are first-order Jahn–Teller unstable and the molecule will distort away from this structure. Structures with four short and two long trans Cu—X bonds are frequently found (Structure **14.1**) but square planar geometries where the two trans ligands have been completely lost are also known (Structure **14.5**). With very few exceptions all low-spin d^8 complexes are square planar (and thus probably associated therefore with a larger driving force for ligand loss than the d^9 case). So these molecules distort in accord with the symmetry principles of the approach. We do though have to be a little careful with this octahedral d^8 case and will come back to it later.

What does the theorem tell us about the magnitude of the distortion?

This is a purely symmetry-based result. As we noted earlier, the theorem tells us nothing about the magnitude of the energy terms involved. It will be a balance between the

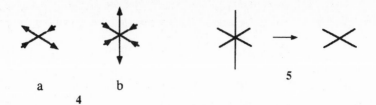

a   b   5

4

energetic preferences of those valence orbitals that lead to the electronic degeneracy (and thus favor the geometric instability) and the filled orbitals of the underlying structure, just as we saw in Chapter 12.

What happens if the distortion energy is small?

It depends what you mean by small, but if the Jahn–Teller energy is comparable to kT (T is the temperature), then instead of a static distortion, a dynamic effect is found where the molecule undergoes large amplitude vibrations about the high symmetry structure. These cases are invariably found when the orbitals involved in the Jahn–Teller instability are non-bonding ones or are involved in weak π-type interactions with the ligands. An example is the molecule $V(CO)_6$ with the $(t_{2g})^5$ configuration. Static Jahn–Teller distortions are found when the molecular orbitals involved are energetically more important, for example, orbitals involved in metal–ligand σ interactions. As we have just mentioned, the Jahn–Teller distortion energy needs to overcome the generally conservative forces associated with the filled orbitals that usually prefer the symmetrical geometry. There are also problems with our simple theory in such cases. The separation of electronic and nuclear motion (the Born-Oppenheimer approximation) is often a problem when the molecule moves through these high-symmetry points with degenerate electronic states.

How can the distortion be viewed in terms of the molecular orbital structure?

Figure 14.2 shows a molecular orbital diagram for the distortion of Structure **14.1** and for one of the opposite phase. Structure **14.6** shows the removal of the degenerate electronic state on distortion for the d^9 configuration. You can see from Figure 14.2 how the identity of the lower energy state is determined by the phase of the distortion through the relevant level ordering. An octahedral d^{10} system is electronically a closed shell and therefore has an $^1A_{1g}$ ground state. From Figure 14.2 one can clearly see that there is no energy gain from a first-order Jahn–Teller distortion. The same result comes from symmetry. With an $^1A_{1g}$ electronic state the first-order Jahn–Teller mode, q, is of a_{1g} symmetry. Since this motion does not change the point group, such a species (an octahedral Hg(II) system, for example) is first-order Jahn–Teller inactive or first-order Jahn–Teller stable.

How does the orbital picture allow access to the d^8 case? We do know that for many molecules (such as $Ni(H_2O)_6^{2+}$) the Jahn–Teller stable octahedral high-spin $^3A_{2g}$ state is found and not the square-planar low-spin variant. Presumably here we have to balance the stabilizing effect of the orbitals (and the electrons in them) in the low-spin state with the energetically destabilizing effect of electron pairing.

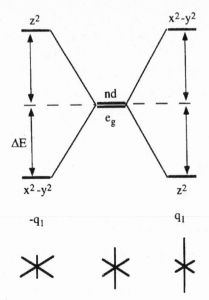

Figure 14.2
A molecular orbital diagram for the distortion of Structure **14.1** *(q_1) and for one of the opposite phase*

Precisely. Figure 14.3 shows two possible scenarios. Figure 14.3(a) shows the case where the electron–electron interactions are high and/or the orbital driving force associated with the distortion to square-planar is low. This will correspond typically to a first-row metal atom with weak-field ligands. Here the $^3A_{2g}$–^1E_g separation is large at the octahedral structure as a result of strong electron–electron interactions. For this case the high-spin, octahedral complex is the one that will be found. Figure 14.3(b) shows the opposite situation, corresponding to a second- or third-row transition metal. With a smaller $^3A_{2g}$–^1E_g separation the square-planar structure results. Thus our comments concerning the Jahn–Teller instability of d^8 have to be qualified a little, albeit in a straightforward way.

What about the second-order term in Equation (14.4)?

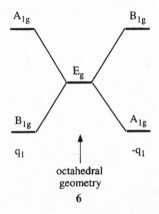

6

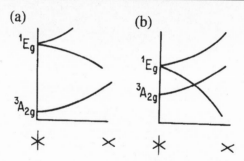

Figure 14.3

Energetic changes of orbitals and states from octahedral to square- planar for the d^8 case: (a) the $^3A_{2g}-^1E_g$ separation is large at the octahedral structure and the orbital driving force octahedral → square-planar is low; (b) the $^3A_{2g}-^1E_g$ separation is small at the octahedral structure and the orbital driving force octahedral → square-planar is high

$E_0^2(q)$ is the second-order correction to the energy and it consists of two terms. Since it depends on q^2 it has the appearance of a Hooke's law force constant. The first term, $\langle 0|\partial^2\mathcal{H}/\partial q^2|0\rangle$, is positive and may be regarded as the force constant that describes the motion of the nuclei in the electronic state $|0\rangle$, namely within the electronic charge distribution of the undistorted $(q=0)$ structure. The second term, $\Sigma'_j|\langle 0|\partial\mathcal{H}/\partial q|j\rangle|^2/(E_j(0)-E_0(0))$, which comes from second-order perturbation theory, is always stabilizing and can be regarded as describing the energy change associated with the relaxation of this charge distribution on motion along the coordinate q. The sign of $E_0^2(q)$ is set therefore, by the sum of these two terms in a similar fashion to that described earlier for the first-order effect. If positive, the molecule is stable and may vibrate around its equilibrium position $q=0$, if negative the molecule will distort along the coordinate q. The largest contributions in this second-order sum will be for electronic states that lie close to the ground state, since they are weighted most heavily from the energy gap expression $E_j(0)-E_0(0)$ which appears in the denominator.

If one such low-lying state, $|j\rangle$, is of the correct symmetry ($\Gamma_q=\Gamma_0\otimes\Gamma_j$) then the first term in this sum can become energetically very important so that $E_0^2(q)$ may become negative. So, in this case the molecule will distort away from the reference geometry.

Yes, there are two different applications of this result, the pseudo- and second-order Jahn–Teller effects. We'll start with the pseudo-Jahn–Teller effect. Our discussion of the first-order Jahn–Teller result showed the importance of a degenerate electronic state for electronic instability as in the ^2E_g state of octahedral Cu(II) or ^1E_g state of octahedral d^8 low-spin Ni(II) with e_g^3 and e_g^2 electron configurations respectively. These are two cases where there is an asymmetric occupation of degenerate orbitals. However, as you pointed out, a degenerate electronic state does not arise from the e_g^2 configuration of square-planar cyclobutadiene. Here (Structure **14.7**) a trio of singlet states, $^1B_{2g}$, $^1A_{1g}$, $^1B_{1g}$, lying energetically in this order, result from the symmetric direct product of e_g. Thus the first-order Jahn–Teller term is zero since none of these states is degenerate. This turns out to be a general result for molecules belonging to the D_{4h} or D_{8h} point groups. The second-order term however, is non-zero by symmetry. If we write $|0\rangle=^1B_{2g}$ and $|j\rangle=^1A_{1g}$ then the symmetry species of the pseudo-Jahn–Teller mode is $\Gamma_q=b_{2g}$. This motion (Structure **14.8**) takes the square to the rectangular structure. This distortion, which arises from the

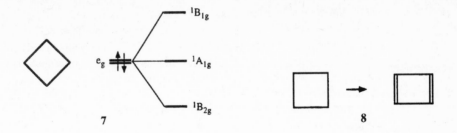

$^1B_{1g}$

e_g $^1A_{1g}$

$^1B_{2g}$

7   8

presence of energetically close, non-degenerate states, which are themselves derived from the asymmetric occupation of degenerate orbitals, is termed a pseudo-Jahn–Teller distortion.

But we can envisage a situation where close-lying states arise from different electronic configurations.

A distortion that we formally call a second-order Jahn–Teller distortion may arise too through a non-zero second-order term. Here though, as you anticipated, the orbital occupation is not associated with degenerate levels and therefore definitely does not give rise to a degenerate electronic state. This is the case for the 'Jahn–Teller' distortion of pseudo-octahedral Cu(II) complexes where the ligands are different ($CuX_4X'_2$) which you mentioned earlier (Structure **14.2**). Now the point group of the undistorted molecule is lowered from O_h to D_{4h} and the doubly degenerate e_g level splits into two. The left-hand side of Structure **14.9** shows the $^2A_{1g}$ and $^2B_{1g}$ states which result. The second-order Jahn–

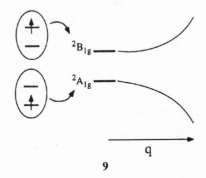

$^2B_{1g}$

$^2A_{1g}$

q

9

Teller mode is of $a_{1g} \otimes b_{1g} = b_{1g}$ symmetry that leads to the observed distortion of the molecule. So the molecule does distort, just as you might expect from the loosest usage of the Jahn–Teller ideas.

Presumably there is a simple explanation for the exclusion of linear molecules from first-order Jahn–Teller considerations? However, they do often distort. NO_2, if it were linear, would have a degenerate electronic state, yet it is a non-linear molecule.

Yes, we can readily show[5] show that for $\Pi_{g,u}$ states of linear molecules the symmetric direct product is $\sigma_g^+ + \delta_g$ and for $\Delta_{g,u}$ states that the direct product is $\sigma_g^+ + \phi_g$, etc. All of the bending modes of linear molecules are of $\pi_{g,u}$ symmetry and thus there will be no first-order stabilization on bending for any (doubly) degenerate state. However, on bending the molecule, the doubly degenerate Π or Δ state will split apart in energy. Now there will be two states separated in energy by a small energy gap. By symmetry they may interact with each other via the quadratic term in the distortion (technically a second-order Jahn–Teller interaction). The overall result is that the system will behave in a similar fashion (Figure 14.1) to that found for the first-order Jahn–Teller effect. This is known[6] as the Renner effect or the Renner–Teller effect.

> Presumably we can use these second-order ideas to look at many other molecular geometry questions?

Yes, but we will leave this until Chapter 17. An example though is shown in Figure 14.4. Imagine an ammonia molecule, forced to be planar in its electronic ground state. The HOMO is of a_2'' symmetry and the LUMO is of a_1' symmetry. The second-order Jahn–Teller mode is of $a_2'' \otimes a_1' = a_2''$ symmetry that leads to bending of the molecule therefore along the a_2'' distortion coordinate. This corresponds to an out-of-plane bending motion that results in the observed pyramidal structure. The result is a stabilization of the lower state or orbital and a destabilization of the upper. Notice how the lone pair ($2a_1$) has developed during the geometry change.

> Yes, but you have suddenly moved from a description based on electronic states to one based on orbitals. I can see it simplifies the problem greatly but is it correct?

The symmetry species of the states $|0\rangle$ and $|j\rangle$ are determined of course by the symmetry of the orbitals that are occupied. We can write the two state wavefunctions as antisymmetrized products (see Chapter 3). For the ground state this will be $|\phi_1\bar{\phi}_1\phi_2\bar{\phi}_2 \cdots \phi_{HOMO}\bar{\phi}_{HOMO}|$ where the bar over the orbital represents an electron of opposite spin. The first exited state may similarly be written $|\phi_1\bar{\phi}_1\phi_2\bar{\phi}_2 \cdots \phi_{HOMO}\bar{\phi}_{LUMO}|$. In the evaluation of the second-order Jahn–Teller term for symmetry purposes the electronic situation reduces to an orbital product, $\phi_{HOMO}\phi_{LUMO}$ and you can see that Γ_q is then just given by the product {symmetry of the orbital holding the electron in the ground electronic state} $\otimes$ {symmetry of the orbital holding the electron in the excited electronic state}. The problem has now been reduced to an orbital rather than state description.

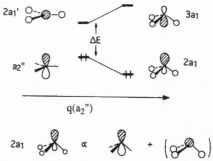

Figure 14.4
The second-order Jahn–Teller effect in planar ammonia

One of the problems associated with your discussion of the first-order Jahn–Teller effect is that the result is not unambiguous in terms of the details of the distorted structure. Given the flexibility allowed by quantum mechanics in writing degenerate wavefunctions, there are an infinite number of possibilities available for writing the degenerate, e_g, Jahn–Teller distortion coordinate for the case of Cu(II) for example. You showed one pair in Structure **14.4**, but, of course any linear combination of the two is also an equally valid choice. Although one of those shown in Structure **14.4** takes the octahedral geometry with six equal distances to one with four short and two long distances, the arrangement found for almost all d^9 systems with six identical ligands, others are equally possible by symmetry, including the opposite phase of Structure **14.4b** (Structure **14.10**) where there are two short and four long distances. What really favors the two-long four-short structure?

10

Figure 14.5 shows the essence of your question. This is the energy of the molecule drawn as a function of the two distortion coordinates q_1 and q_2 of Structure **14.4(b), (a)**. It is often called a 'Mexican hat' potential since the surface is equi-energetic for a given $|q|$ for all ϕ using the generalized coordinate $q = q_1 \cos \phi + q_2 \sin \phi$. That this might be the case is seen in Figure 14.2 which shows how the d orbitals change in energy during the distortion for the two cases $\pm q_1$. Energetically they are identical, the only difference between the two is the identity of the upper and lower orbitals. Although there are suggestions that the stabilization of one distortion mode over another comes from cooperative effects in the crystal (where most Jahn–Teller distortions have been found), this cannot be the whole picture. Both $CuBr_2$ and $HgBr_2$ adopt a distorted form of one of the cadmium halide structures. In one the distortion results in a two-long four-short structure (Cu) and in the other (Hg) a four-long two-short arrangement.

Presumably we need to go outside the d orbital manifold to understand this?

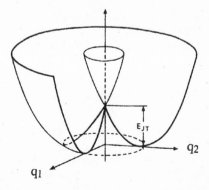

Figure 14.5
*The 'Mexican hat' potential; the energy of the Jahn–Teller unstable molecule drawn as a function of the two distortion coordinates q_1 and q_2 of Structures **14.4b** and **14.10***

Yes, the clearest picture in fact comes from consideration[2,7] of the valence $(n+1)$s orbitals (4s in the copper case) in addition to those of nd. The energy separation between these two sets of orbitals may well be small, especially at the right-hand side of the Periodic Table. We are also cognizant of the general importance of metal $(n+1)$s, p orbitals in transition-metal chemistry in general. This leads to a small energy denominator, $(E_j(0) - E_0(0))$ associated with the $(n+1)$s/nz^2 separation (Structure **14.11**). At the octahedral geometry the $(n+1)$s orbital has a_{1g} symmetry, the nz^2, $n(x^2 - y^2)$ e_g symmetry, and thus a second-order interaction between $(n+1)$s and nd is allowed by symmetry along the distortion coordinate of e_g symmetry that takes the octahedral structure to D_{4h}. However, there is a catch here. The $(n+1)$s orbital can only interact by symmetry with one of the e_g components along the distortion coordinate q. (Remember $e_g \rightarrow a_{1g} + b_{1g}$ for $O_h \rightarrow D_{4h}$.) On distortion we can see (Figure 14.6) that the $(n+1)$s orbital may interact with nz^2 (they both have a_{1g} symmetry) but not with the $x^2 - y^2$ orbital (b_{1g}) in the D_{4h} point group.

These $(n+1)$s/nz^2 orbital interactions (Figure 14.6) show in a straightforward way why the two-long four-short distortion (q_1 of Figure 14.2) is found for the d^8 or d^9

$$(n+1)s \;\; \text{———} \;\; a_{1g}$$

$$nd \;\; \text{═══} \;\; e_g$$

11

configuration. If we put the gap, $(E_j(0) - E_0(0))$, of equation (14.4) equal to the $(n+1)$s/z^2 separation, the stabilization of z^2 will be larger in the two-short four-long case than for the converse ($|\Delta E_1| < |\Delta E_2|$).

> This comes about because the z^2 orbital is pushed up in energy in the two-short four-long geometry resulting in a smaller energy gap between z^2 and s than for the distortion of the opposite phase.

But the electron occupancy is important. In the low-spin d^8 case only one of the e_g pair of orbitals is occupied. Thus the geometry found will be the one where the d orbital that is stabilized by interaction with the $(n+1)$s orbital is the one which is occupied. A set of energy differences for the two pathways is shown in Table 14.1 for the d^8–d^{10} configurations. $|\Delta|$ is the energy change of either z^2 or $x^2 - y^2$ on distortion using the d orbital only, model. For both d^8 and d^9 the two-long four-short geometry is the one predicted and found experimentally. Notice the larger distortion energy for low-spin d^8 compared to d^9 systems, a result in accord with experiment. The square planar is the structure almost universally found for the low-spin d^8 case.

> And a similar argument is applicable to closed-shell d^{10} systems where there is no first-order effect possible?

Yes, a similar argument shows why the two-short four-long distortion ($+q_1$) is found for d^{10} systems such as those of Hg(II) where there is no first-order Jahn–Teller contribution.)

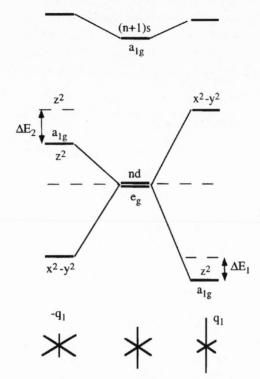

Figure 14.6
Modification of the diagram of Figure 14.2 that shows how inclusion of $(n+1)s/nz^2$ orbital interactions lead to prediction of the two-long four-short distortion $(-q_1)$ is found for the d^8 or d^9 configuration but the opposite for d^{10}

In the d^{10} configuration (Table 14.1) the best electronic stabilization occurs for the two-short four-long case because nz^2 lies closer to $(n+1)s$. The tendency for Hg(II) compounds to adopt linear two-coordinate structures has long been interpreted[8] in terms of strong 5d/6s mixing. Here is the same idea from a different viewpoint.

There are a couple of additional points here though. You have focused on the elegant symmetry and orbital occupation aspects of the approach à la Jahn and Teller. From the database of distorted Cu(II) systems, as far as I can see, if all of the ligands are the same then this two-long four-short structure is the one found, although sometimes in solids the local

Table 14.1 Energetic stabilization[a] on distortion from Figure 14.6

Configuration	q_1	$-q_1$
d^7	$\Delta + \Delta E_1{}^*$	Δ
lsd^8	$2\Delta + 2\Delta E_1{}^*$	2Δ
d^9	$\Delta + 2\Delta E_1{}^*$	$\Delta + \Delta E_2$
d^{10}	$2\Delta E_1$	$2\Delta E_2{}^*$

[a] The more stable structure is labeled with an *.

environment may force the opposite distortion. However, in complexes where the ligands are different, these symmetry arguments are sometimes not so useful. So in the $MY_4Y'_2$ complex of Structure **14.12** (with $2Y + 2Y'$ in the plane perpendicular to the long metal–ligand distances) the distortion is of b_{1g} symmetry as expected from the symmetry requirements of the second-order Jahn–Teller effect for the D_{4h} complex. However, in Structure **14.13** (with $4Y$ in this plane), a structure type found for many CuO_2N_4 complexes of Cu(II) (O and N here represent a range of oxygen and nitrogen donors), the distortion mode is clearly a_{1g}, which is not in accord with such arguments.

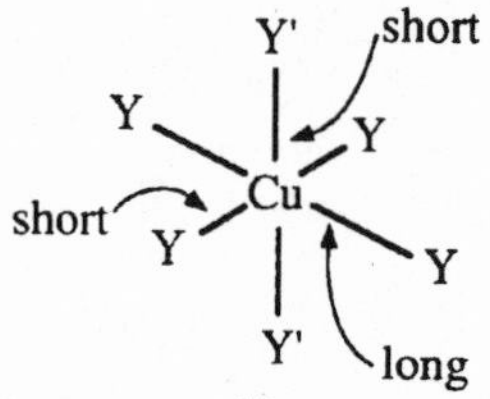

12 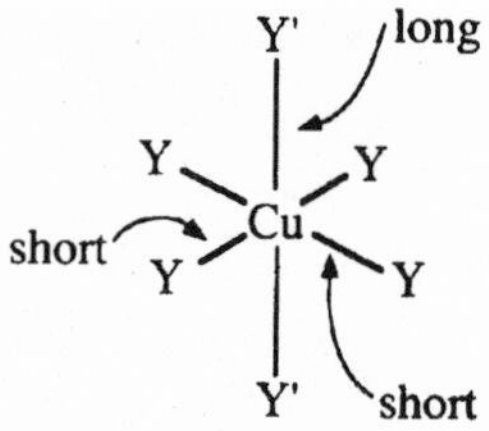   13

This is indeed a limitation of the symmetry approach, but one that is readily understood[9]. Let's use the AOM of Chapter 7 to calculate the energies of the two distorted structures, (**14.12**) and (**14.13**). From Structures **14.14** and **14.15** the MOSE for the two possibilities are $(3/2)(e_\sigma(Y) + e_\sigma(Y'))$ and $3e_\sigma(Y)$ respectively. Thus Structure **14.12** will be favored over **14.13** when $e_\sigma(Y) < e_\sigma(Y')$. If nitrogen donors have larger values of $e_\sigma(Y) + e_\sigma(Y')$ than oxygen donors (as they do) then the structural results of Structures **14.12** and **14.13** are readily accessible. So for Structure **14.12** we may have the right result for the wrong reason from the Jahn–Teller approach. There is a sobering comment[10] by J. H. van Vleck that runs as follows: 'A really good theoretical chemist can obtain the right answers with wrong models'.

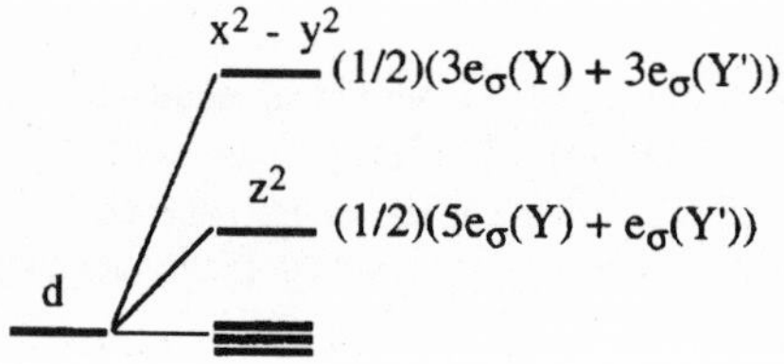

14

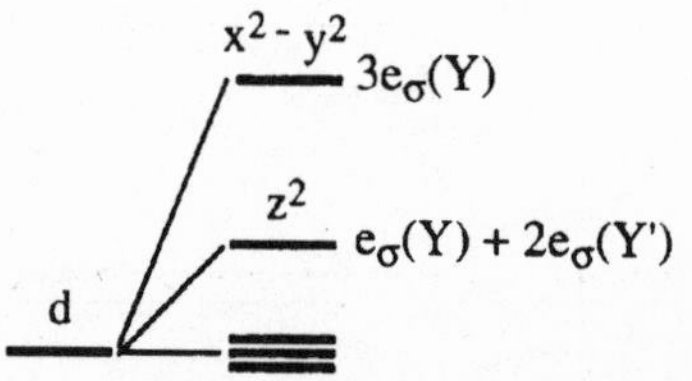

15

Most of our first-order Jahn–Teller examples have been transition metal- containing ones. This is probably because the high symmetry of many such complexes and the fact that there are five d orbitals, may frequently lead to degenerate orbitals. But there must be some illustrative main group cases.

Quite true. One particularly nice example is the demonstration that B_2H_6 (Structure **14.16**) cannot have the same structure as C_2H_6 (Structure **14.17**). Figure 14.7 shows a part of the molecular orbital diagram for ethane. Notice that the HOMO (e_g) of staggered ethane is a doubly degenerate orbital of π type largely involved in C–H σ-bonding. With two fewer electrons, the ethane structure is now first-order Jahn–Teller unstable and B_2H_6 certainly will not exist in this structure. There is a nice orbital correlation diagram coupling the two structures in reference 11.

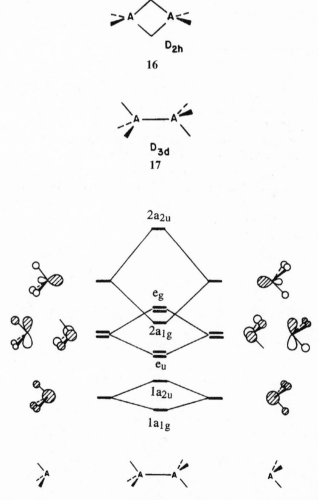

Figure 14.7
Assembly of a part of the molecular orbital diagram for ethane. Notice that the HOMO (e_g) is doubly degenerate

References

1. Jahn, H. A. and Teller, E., *Proc. Roy. Soc.*, **A161**, 220 (1937).
2. Burdett, J. K., *Molecular Shapes: Theoretical Models of Inorganic Stereochemistry*, John Wiley & Sons (1980).
3. Bartell, L. S., *J. Chem. Educ.*, **45**, 754 (1968).
4. Bersuker, I. B., *The Jahn–Teller Effect and Vibronic Interactions in Modern Chemistry*, Plenum (1984).
5. Albright, T. A. and Burdett, J. K., *Problems in Molecular Orbital Theory*, Oxford University Press (1992).
6. Renner, R., *Z. Phys.*, **92**, 172 (1934).
7. Gerloch, M., *Inorg. Chem.*, **20**, 638 (1981).
8. Dunitz, J. D. and Orgel, L. E., *Adv. Inorg. Chem. Radiochem.*, **2**, 1 (1960).
9. Burdett, J. K., *Inorg. Chem.*, **20**, 1959 (1981).
10. Van Vleck, J. H., *22nd Annual Congress of Pure and Applied Chemistry; Plenary Lectures*, Butterworth (1970).
11. Albright, T. A., Burdett, J. K. and Whangbo, M.-H., *Orbital Interactions in Chemistry*, John Wiley & Sons (1985).

General References and Further Reading

Bader, R. F. W. and Bandrauk, A. D., *J. Chem. Phys.*, **49**, 1666 (1968).
Bersuker, I. B., *Electronic Structure and Properties of Transition Metal Compounds*, John Wiley & Sons (1996).
Englman, R., *The Jahn–Teller Effect in Molecules and Crystals*, John Wiley & Sons (1972).
Öpik, U. and Pryce, M. H. L., *Proc. Roy. Soc.*, **A238**, 425 (1957).
Pearson, R. G., *Symmetry Rules for Chemical Reactions; Orbital Topology and Elementary Processes*, John Wiley & Sons (1976).

15 What is the Origin of Steric Repulsion?

Virtually all of our attention has been with questions of chemical bonding, but what about those 'antibonding forces', those that keep atoms and molecules apart. Is there a simple way to visualize them?

Obviously at short range, nuclei begin to repel each other strongly as do ions charged in like manner, and simple mathematical forms for both ionic and van der Waals interactions formally contain short-range repulsive terms in the manner in which they are most generally used. The ionic model utilizes an interatomic potential of the form $z_1z_2e^2/r + A/r^n$ for two charges, z_1, z_2 separated by a distance r. The lead, electrostatic term, is balanced (when attractive) by the second repulsive term. (n, the Born exponent, depends on the closed-shell configurations of the ions.) We have used a $1/r^n$ interaction here but an exponential form is often used. Van der Waals interactions are commonly mimicked by a potential of the form $-B/r^6 + C/r^{12}$, again one containing a short-range repulsion. Some simple orbital models add a repulsive term to accommodate inadequacies in computing correct bond lengths, but the orbital model itself naturally leads to a repulsion arising from orbital orthogonalization for closed shells of molecules.

The H_2 and He_2 molecules will serve as useful models. In order to obtain their energy levels we need to solve a secular determinant ($|H_{ij} - S_{ij}E| = 0$) which for these simple diatomics involves the two 1s orbitals on the two atoms (ϕ_1, ϕ_2)

$$\begin{vmatrix} H_{11} - E & H_{12} - S_{12}E \\ H_{21} - S_{21}E & H_{22} - E \end{vmatrix} = 0 \tag{15.1}$$

or using the usual labels of Hückel theory,

$$\begin{vmatrix} \alpha - E & \beta - SE \\ \beta - SE & \alpha - E \end{vmatrix} = 0 \tag{15.2}$$

which gives

$$E = (\alpha + \beta)/(1 + S) \quad \text{and} \quad E = (\alpha - \beta)/(1 - S) \tag{15.3}$$

The corresponding wavefunctions are given by

$$\psi_b = 1/\sqrt{2(1 + S)}(\phi_1 + \phi_2)$$
$$\psi_a = 1/\sqrt{2(1 - S)}(\phi_1 - \phi_2) \tag{15.4}$$

These molecular orbitals were derived in a different way by simply using symmetry principles in Chapter 1. The important aspect of their derivation here, is that overlap has been included in the normalization process. A molecular orbital diagram based on equations (15.2) and (15.3) is shown in Figure 15.1. The vital point here is that the bonding orbital has been stabilized less than the anti-bonding orbital has been de-stabilized. This simple fact leads to a ready understanding of why the He_2 molecule does not exist. Of the four electrons, two are stabilized on formation of the molecular orbitals (as in H_2) but the other two are destabilized and by a larger amount. You can do a little

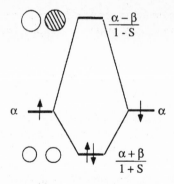

Figure 15.1

A molecular orbital diagram for H₂ derived using equations (15.2) and (15.3) showing that the bonding orbital is stabilized less than the anti-bonding orbital is destabilized

arithmetical manipulation to show that its energy of formation is approximately $-4\beta S/(1 - S^2)$. $\beta < 0$ so this represents a destabilization of the system. Such effects are sometimes called four-electron two-orbital repulsions and sometimes described as due to Pauli repulsion since by forcing two electrons to occupy a bonding orbital and two to occupy an anti-bonding orbital (required by the Pauli Principle) the overall effect is a de-stabilizing one.

> Since the overlap integral, S, generally increases in magnitude as the interatomic separation decreases, presumably the repulsive effect is amplified as the internuclear separation becomes shorter.

Of course, and at really short distances the nuclear–nuclear repulsion becomes huge, but these distances are not those of interest to the chemist.

There is another way to look at the bonding propensities between two atoms by a study of the electron density distribution in the molecule[1]. Let's take a general wavefunction given by equation (15.5):

$$\psi = a\phi_1 + b\phi_2 \tag{15.5}$$

This has an electron density distribution function

$$\psi^2 = a^2\phi_1^2 + b^2\phi_2^2 + 2ab\phi_1\phi_2 \tag{15.6}$$

Notice that some of the electron density is located on ϕ_1, some on ϕ_2 and some shared between the two. Multiplication by N, the number of electrons in the orbital and integrating gives

$$\begin{aligned} N\langle\psi^2\rangle &= Na^2\langle\phi_1^2\rangle + Nb^2\langle\phi_2^2\rangle + 2Nab\langle\phi_1|\phi_2\rangle \\ &= Na^2 + Nb^2 + 2NabS_{12} \end{aligned} \tag{15.7}$$

This final expression shows that there are Na^2 electrons located on ϕ_1, Nb^2 electrons located on ϕ_2 $2NabS_{12}$ electrons located in the bond between the two. This last term is a very interesting one. $2NabS_{12}$ is the contribution to the Mulliken bond overlap population[2] between the two atoms via occupation of this orbital and is a useful way to quantify chemical bonding by occupation of a given energy level.

Presumably the integrated (for solids) or summed (for molecules) orbital overlap populations over all the occupied orbitals gives a measure of the total bond strength between a given pair of atoms?

Yes. The total bond overlap population between a pair of atoms a, b in a molecule is thus

$$P_{mn} = 2 \sum_i N_i c_{ij} c_{ik} S_{jk} \tag{15.8}$$

Here c_{ij} is the coefficient of orbital j located on atom m, c_{ik} the coefficient of orbital k located on atom n in molecular orbital i, and S_{jk} is their overlap integral. N_i is the number of electrons in the ith molecular orbital. A positive P_{mn} represents a bonding situation, a negative value an anti-bonding one. Let's evaluate the bond overlap population between the two hydrogen atoms in H_2 for the configuration ψ_b^2. From the form of the wavefunction in equation (15.4),

$$\begin{aligned} P_{12} &= 2.2.1/\sqrt{2}(1+S) \cdot 1/\sqrt{2}(1+S).S \\ &= 2S/(1+S) \end{aligned} \tag{15.9}$$

and for the configuration ψ_a^2 it is analogously

$$-2S/(1-S) \tag{15.10}$$

So in accord with energetics associated with the diagram of Figure 15.1 the anti-bonding orbital is more anti-bonding ($P_{12} < 0$) than the bonding orbital is bonding ($P_{12} > 0$). The total overlap population for He_2 with the configuration $\psi_a^2 \psi_b^2$ is thus

$$-4S^2/(1-S^2) \tag{15.11}$$

by adding the results of equations (15.9) and (15.10). This is negative and represents an anti-bonding situation. Thus in a different way we can see how a repulsion arises between atoms. Notice that it comes about through the overlap term in the orbital normalization. The same ideas are just as useful in more complex molecules. One has to be careful though in comparing the magnitudes of bond overlap populations between different atom pairs. Thus one should not compare the strength of bonding or anti-bonding interactions in Mn–Mn linkages and C–C linkages, for example, by using computed numerical values.

Although it is easy to see that there is a bonding interaction between the central carbon atom and the peripheral hydrogen atoms in methane, what about those between the hydrogen atoms themselves? We certainly draw the structure of the molecule with four C–H bonds and ignore interactions between the latter.

Figure 15.2 shows the hydrogen atom wavefunctions of a_1 and t_2 symmetry which interact with the carbon 2s and 2p orbitals. Assuming that the interactions of these two sets of orbitals are similar (i.e., $p \sim q$) we can just use the simple form of these molecular orbitals to evaluate the bond overlap populations between the hydrogen atoms in the methane molecule. We need to sum (equation (15.8)) over all occupied orbitals. S' is the overlap integral between a pair of hydrogen 1s orbitals:

For a_1: $P_{HH} = 2.2.(1/2).(1/2)(1/\sqrt{(1+3S)})^2 p^2 S' = p^2(1/(1+3S))S'$

For t_2: $P_{H1H2} = 2.2.(3/\sqrt{12}).(-1/\sqrt{12})(1/\sqrt{(1-S)})^2 p^2 S' = -p^2(1/(1-S))S'$

$$\tag{15.2}$$

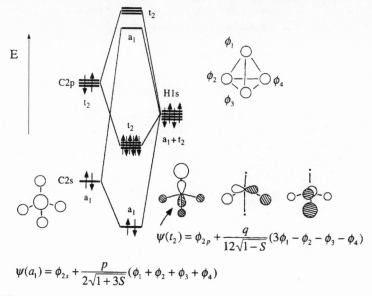

$$\psi(t_2) = \phi_{2p} + \frac{q}{12\sqrt{1-S}}(3\phi_1 - \phi_2 - \phi_3 - \phi_4)$$

$$\psi(a_1) = \phi_{2s} + \frac{p}{2\sqrt{1+3S}}(\phi_1 + \phi_2 + \phi_3 + \phi_4)$$

Figure 15.2
Form of the wavefunctions of a_1 and t_2 symmetry in methane (only one of the t_2 functions is described algebraically)

From our choice of the wavefunctions (recall that for a degenerate set of wavefunctions there are an infinite set of ways of drawing them, equally as valid), there is only one t_2 component that makes a contribution to the 1–2 overlap population (which is why we chose this particular member of the set for our analysis). You can do the same calculation for 1–3, 2–3 atom pairs etc., and get the same answer (as you have to by symmetry). And the total overlap population is

$$-4S/(1+3S)(1-S).p^2S' \tag{15.13}$$

which is negative and thus represents a repulsion between the hydrogen atoms.

It's interesting to note that if we neglect overlap in the normalization then these repulsions disappear. This gives us a nice way to view them. Overlap leads to chemical bonding if the orbital occupation is right but the opposite if it isn't. Are things always as simple as this?

Our discussion here has used two basic concepts. The first is the expansion of the molecular wavefunction as a linear combination of the valence orbitals on the constituent atoms, and the second the conceptually simple idea of the Mulliken bond overlap population. In high quality calculations of the *ab initio* type the wavefunction is much more complex and involves a much larger basis set. Under these conditions the reliability of the Mulliken population analysis is open to question. It is an unfortunate but common feature of quantum mechanical calculations that as they become more accurate they become more difficult to understand. Recall my comment of Robert Mulliken's in the Prologue.

We've talked about non-bonded repulsions between 'non-bonded atoms' but presumably there are repulsions of the same type between bonded atoms if the orbital occupancy is appropriate.

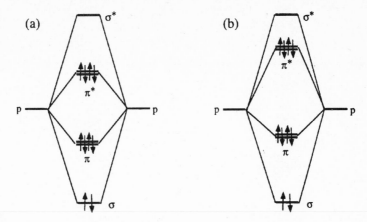

Figure 15.3
Molecular orbital diagrams for a main group diatomic molecule ignoring the valence s orbitals for simplicity: (a) shows the case where pπ interactions are not very important; and (b) the case where they are

Yes, the classic case is that for the fluorine molecule. The bond dissociation energy of the halogens varies down the series in the following way. F_2 (159), Cl_2 (243), Br_2 (193), I_2 (151 kJ/mol). Bond energies normally decrease on going down a group but it's very interesting to note that the bond dissociation energy for F_2 is much smaller than we would expect by extrapolation, and is close to that for I_2, the heaviest of the series. Figure 15.3 shows a molecular orbital diagram for a diatomic halogen molecule that ignores the valence ns orbitals for simplicity (although they are beginning to become core-like at this point in the Periodic Table). Notice that the configuration ($\sigma_g^2 \pi_u^4 \pi_g^4$) leads to occupation of both π bonding and anti-bonding orbital sets. If π interactions are important as they are for first-row atoms (see Chapter 4) then there will be a strong repulsion between the two atoms from this effect. In other words the bond strength anticipated by the presence of the σ bond is offset by these two four-electron two-orbital repulsions. A similar effect is found for the single-bond energies involving first-row atoms in general. For example the bond energies in the two A–O series (A = Group 14, 15 elements), N–O (157), P–O (368), As–O (330 kJ/mol) and C–Cl (338), Si–Cl (401), Ge–Cl (339) and Sn–Cl (314 kJ/mol). Such energies generally decrease down a group but those for the pair of first-row atoms are anomalously weak.

References

1. Albright, T. A., Burdett, J. K. and Whangbo, M.-H., *Orbital Interactions in Chemistry*, John Wiley & Sons (1985).
2. Mulliken, R. S., *J. Chem. Phys.*, **23**, 1841 (1955).

16 Why does N_2O have the Structure NNO and not NON?

One of the features of structural chemistry is that many molecules and solids have 'symmetrical' structures. So it's a little surprising that N_2O has the asymmetric arrangement NNO where the oxygen atom lies at the end of the triatomic molecule and not in the middle. This is not the only example. The trihalide ions always have asymmetric structures, $ClClF^-$ for example. Also there are changes in connectivity with isoelectronic systems. So whereas FNO has the connectivity implied by the way I have written its formula, FNS has the arrangement FSN. There does seem to be a rule. Place the most electronegative atom at the end of the molecule. Is this a general picture, and if so what is its origin?

The rule is, in fact, of general validity for many molecules. As always it's controlled by electron count. It's a result of topological charge stabilization[1], in other words a function of the way the atomic connectivity determines the way charge is distributed over the molecule. The general idea is a simple one. Let's start with the charge distribution in the isoelectronic and isostructural molecule N_3^- where all three atoms are the same. We can readily construct the three sets of π orbitals of the linear triatomic as in Figure 16.1 within the Hückel approximation. It is easy to write down the orbital coefficients as shown. The HOMO of the molecule is the middle pair of the set, π_n, the non-bonding orbital. The electron densities in the Hückel sense are $4((1/2)^2 + (1/\sqrt{2})^2) = 3$ for the terminal and $4((1/\sqrt{2})^2) = 2$ for the middle atom of the molecule using the electron configuration $(\pi_b^4\pi_n^4)$ appropriate for the sixteen-electron molecule. Since each atom contributes $8/3 = 2.67$ electrons in this barebones diagram, the central atom has a charge of $+0.67$ and the terminal atoms a charge of -0.33 electrons each, relative to $N^{-0.333}$.

Now we can use these results to see where the oxygen atom prefers to reside in N_2O by using the results of perturbation theory. In first order at least, this is best viewed in the following way. Let's write an LCAO expression for the jth molecular orbital in the usual way as

$$\psi_j = \sum_i c_{ij}\phi_i \tag{16.1}$$

Now, using the expression for the energy, $E_j = \langle \psi_j | \mathcal{H} | \psi_j \rangle$,

$$E_j = \left\langle \sum_i c_{ij}\phi_i | \mathcal{H} | \sum_i c_{ij}\phi_i \right\rangle$$
$$= \sum_i \sum_k c_{ij}c_{kj} \langle \phi_i | \mathcal{H} | \phi_k \rangle \tag{16.2}$$

Separating the two types of terms in this expression and using the Hückel labels α and β,

$$E_j = \sum_j c_{ij}^2 \alpha_i + \sum_i \sum_{k \neq i} c_{ij}c_{kj}\beta_{ik} \tag{16.3}$$

Now let's see what happens when we change α on one of the orbitals ($i = m$).

$$\Delta E_j = c_{mj}^2 \delta\alpha_m + \sum_{k \neq m} c_{mj}c_{kj}\delta\beta_{mk} \tag{16.4}$$

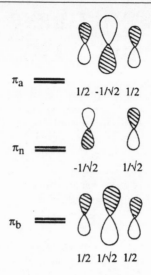

π_a ═══ 1/2 -1/√2 1/2

π_n ═══ -1/√2   1/√2

π_b ═══ 1/2 1/√2 1/2

Figure 16.1
The form of the π levels for a linear triatomic molecule and the atomic orbital coefficients and electron densities from a simple Hückel model. The orbital occupation is $(\pi_b)^4(\pi_n)^4$ for N₂O and $(\pi_b)^4$ for Ga₂O

The total energy change is then

$$\Delta E_{\text{total}} = \sum_j n_j \left(c_{mj}^2 \delta\alpha_m + \sum_{k \neq m} c_{mj} c_{kj} \delta\beta_{mk} \right) \tag{16.5}$$

where n_j is the number of electrons in orbital j. Assuming for simplicity that there is no change in β

$$\Delta E_{\text{total}} = \sum_j n_j \left(c_{mj}^2 \delta\alpha_m \right) \tag{16.6}$$

If there is an unevenness in the charge distribution of the molecule, as will always be the case if there are symmetry inequivalent atoms, then substitution of an atom by one of greater electronegativity $\delta\alpha < 0$, leads to the best stabilization at that site where $\sum_j n_j c_{mj}^2$ is largest. This term is just the electron density we calculated above. So in the case of N₂O, since it is the terminal atoms of the N_3^- structure that have the largest electron density, this is where the more electronegative oxygen atom lies preferentially. The result in general depends on the structure. However, for 'octet' systems it is invariably the site of smallest coordination number that carries the largest charge. So in N₂O the oxygen atom is best stabilized at a terminal position, the site of lowest coordination number.

You can do a similar calculation for the carbonate ion. Here the carbon atom, the least electronegative atom, lies at the site of highest coordination number. Carbonate ion would be a pretty strange species if an oxygen atom lay at the center of the molecule. Your FNO and FNS (FSN) are variants of this rule. Invariably, metal complexes contain a central electropositive atom surrounded by more electronegative ligands. ClF₃ contains a central chlorine, not fluorine, atom. And I'm sure you can think of others, including sulfate ion, nitrate ion, and many of the common species of inorganic chemistry. An example related to FNO/FSN is the structure of realgar (Structure **16.1**) and its nitrogen analog (Structure

●= S   O = As   ●= S   O= N

        1   2

16.2). Notice how the group 15 and 16 atoms have exchanged places, their coordination numbers determined by their relative electronegativities.

> There must be many other structural questions that rely on the same philosophy even though the coordination numbers may be the same for the various sites.

Yes, the idea runs throughout much of chemistry. Figure 16.2 shows a variant of the general question, namely where do the chlorine and methyl ligands lie in the $TeCl_2(CH_3)_2$ molecule? The charges calculated for the fluorine atoms in the prototypical SF_4 molecule, assuming that all of the S–F distances are equal, are different, a result reflecting the different environments for axial and equatorial linkages. Notice the larger charge on the axial fluorine atoms compared to the equatorial ones. So the $TeCl_2(CH_3)_2$ isomer containing an equatorial methyl group is the more stable one, and is the one actually found. It is the one where the more electronegative chlorine atom searches out the site with the largest charge. In the same way the two bromine atoms occupy axial sites and the two phenyl groups, the equatorial sites in $TeBr_2Ph_2$. Similarly in SF_3CH_3 the more electronegative fluorine atoms occupy the axial sites and relegate the methyl group to an equatorial one. In T-shaped ClF_3 the axial sites again carry the highest charge and so in ICl_2Ph, the phenyl group occupies the equatorial site. Exactly analogous arguments place the methyl group in trigonal bi-pyramidal PF_4CH_3 in an equatorial site.

The same considerations apply to cage and cluster molecules. Figure 16.3(a) shows the charges calculated[1] for the cage molecule P_7^{3-}. It immediately leads to an understanding of the structure of P_4S_3. Here the coordination number rule applies. Similar considerations apply to substitution patterns of boron cages although here in this 'electron-deficient'

Figure 16.2

A selection of molecules which show how the most electronegative atoms in a substituted species occupy the sites of highest electron density in the unsubstituted parent. The figure shows computed charges for the terminal atoms of various SF_n molecules, the relevant bond overlap populations and the observed structures of some substituted molecules (adapted from Burdett, J. K., Molecular Shapes: Theoretical Models of Inorganic Stereochemistry, John Wiley & Sons (1980))

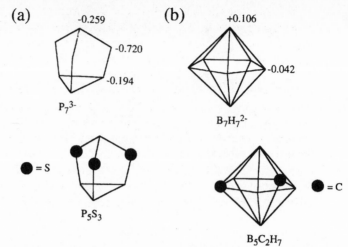

Figure 16.3

(a) The calculated charges in the P_7^{3-} molecule and the observed structure of P_5S_3; (b) the calculated charges in the $B_7H_7^{2-}$ molecule and the observed structure of $B_5C_2H_7$

molecule we might not want to rely on the coordination number rule. Figure 16.3(b) shows, though, that the charges on the equatorial boron atoms (those of lowest coordination number) in $B_7H_7^{2-}$ are indeed larger than on the pair of axial atoms. The 2,4 isomer of $C_2B_5H_7$ is the only one known.

You mentioned that the coordination number–electronegativity rule does depend on electron count. I can't think of any examples.

Let's turn to Figure 16.1 and look at the charge distribution for the case where there are only four π electrons (π_b^4). Now it is the central atom that carries the larger charge. In accord with this result, Ga_2O (a twelve-electron molecule only characterized in the gas phase and certainly not a common species) has the structure GaOGa. B_2O and B_2N, species characterized in matrices, have analogous structures with the electropositive atoms at the end of the molecule. There are examples from the solid state too. Cs_2O has a structure where layers of cesium atoms sandwich a layer of oxygen. With this electron count a calculation on the extended solid shows that it is the more electronegative atom that should have the higher coordination number as implied by our description. There is not really a general rule, but it is electron-precise (octet, eighteen-electron or Wade's rule) systems that most often obey the rule you stated and electron-poor ones which show this reversal.

What happens if the electronegativities of the atoms involved are similar?

The answer to this is obvious. The various isomers will be close in energy. An example is Cl_2O. Its regular structure has the connectivity ClOCl but when isolated in an argon matrix at 20 K, on photolysis the molecule rearranges to ClClO. Species such as NF_2 with a much larger electronegativity difference are only known as FNF.

Are there exceptions to the rules?

Yes, there are. In fact we should be surprised it works as well as it does since there is more than one factor that controls these site preferences. From Figure 16.2, notice that although the most electronegative ligands are predicted to occupy the axial sites in SF_4, for example, the overlap populations associated with the axial linkages are smaller than those for the equatorial ones. So the topological rules will be upset in those cases where the ligand location is set by bond strength rather than by simple electronegativity considerations. In addition, substitution of one atom for another, brings with it, not only a change in H_{ii}, but also a change in overlap integrals with its neighbors. (This will always happen in practice of course since the central atom–ligand distances for the two sites are always different.) Reversal of the topological rules is found[3] in the molecule of Structure **16.3**. (The two unlabeled sites are a part of an eight-membered ring.) In sterically hindered situations the effective 'bond energies' for the two sites may also be quite different and this effect will determine the site preferences. An example might be that[4] of Structure **16.4**.

The cyclic molecule S_2N_2 (Structure **16.5**) also shows a site preference reversal. A simple Hückel calculation on the cyclic molecule A_4 molecule shows that, unlike the system with two pairs of π electrons Structure **16.6** is preferred over the observed Structure **16.5**.

> What happens if I have (say) three heteroatoms to arrange in a molecule where all the atoms are equivalent in the parent?

An example of this question concerns the substitution patterns of organic rings containing more than one heteroatom. So how, for example, will three N atoms arrange themselves when substituted for CH groups in benzene? We can access this problem in a particularly simple way by first constructing a part of the π molecular orbital diagram for pyridine using perturbation theory[2], and then studying the charge distribution in this molecule. Figure 16.4 shows how the HOMO and LUMO of benzene (we show only one component) and a pair of orbitals further apart in energy, mix as a result of the introduction of the electronegative nitrogen substituent. We know that the electron density must increase at the nitrogen site compared to the benzene parent and this tells us the phase of the orbital mixing. Notice that the *ortho* and *para* sites experience a decrease in electron density but the *meta* site an increase. (Obviously the total number of electrons must remain constant.) Thus the most stable arrangement will be the 1,3,5 structure for the tri-aza molecule where the second two nitrogen atoms go into positions *meta* to the first.

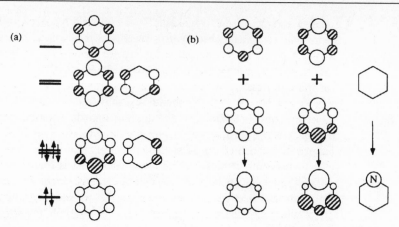

Figure 16.4
Illustration, via perturbation theory, of the change in electron density in a part of the π system of benzene on substitution of a CH group by N. Only the mixing of one component of the degenerate pairs of orbitals is shown. An increase in the orbital coefficients (and thus electron density) at the meta sites is the result

It also tells us why borazine ($B_3N_3H_6$) has the 1,3,5 structure (Structure **16.7**). It is just like the electronegative substitution process of two nitrogen atoms in pyridine (Structure **16.8**). The same is true in Structure **16.9**.

      7   8   9

Yes, it's an interesting illustration of the versatility of the orbital approach used in this book. The same model is able to move smoothly from one area of chemistry to another and, as a consequence, we benefit from the broader picture which results.

References

1. Gimarc, B. M., *J. Amer. Chem. Soc.*, **105**, 1979 (1983).
2. Heilbronner, E. and Bock, H., *The HMO Model and its Applications*, John Wiley & Sons (1976); Imamura, A., *Molec. Phys.*, **15**, 225 (1968).
3. Timosheva, N. V., Prakasha, T. K., Chandrasekaran, A., Day, R. O. and Holmes, R. R., *Inorg. Chem.*, **35**, 4525 (1995).
4. Dunmur, R. E., Murray, M, Schmutzler, R. and Gagnaire, D., *Z. Naturforsch.*, **B25**, 903 (1970).

17   What is Behind the VSEPR Scheme?

One of the sets of rules that are very useful in predicting the geometries of main group molecules are those invented by Sidgwick and Powell and developed by Nyholm and Gillespie. They are known by different names in different parts of the world, either as the Nyholm–Gillespie Rules or as the Valence Shell Electron Repulsion Rules. One simply counts up the total number of σ electron pairs around a central atom, arranges them to minimally repel one another in space (via so-called 'Pauli forces') and then adds the atoms or groups linked to the central one. Any left-over pairs are lone pairs and the gross molecular geometry is thus defined.[1] The geometries we expect are shown in Figure 17.1. Virtually all of our discussions here have involved orbital models but this is a remarkably simple way to approach this question. Is there an orbital picture behind this approach that will allow insights into its operation?

As a rule of thumb the rules work remarkably well. It's an interesting approach since the geometry is set by the repulsion of pairs of electrons. In their original formulation Sidgwick and Powell used the idea of electrostatic repulsion between electron pairs. 'Pauli forces' came later. Molecular orbital methods (of course) also lead to identification of the correct geometries for these molecules. Of interest in this context are the calculations of the one-electron type that are successful in this way (such as those which use the Hückel or extended Hückel methods). These calculations ignore electron–electron interactions and focus on orbital overlap. VSEPR ignores orbital effects and focuses on electron–electron

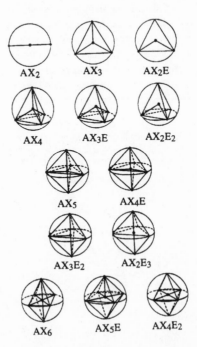

Figure 17.1
The geometries expected using the Valence Shell Electron Repulsion Rules for AX_n

interactions. So you couldn't imagine two models further apart in broad concept. Note though that the orbital model will give you numerical values of the bond angles but the VSEPR scheme, just a qualitative picture of the structure. In addition the orbital model allows access to the geometries of excited states, an area inaccessible to VSEPR.

In VSEPR terms, there are basically two types of compound, those which contain lone pairs (AX_nE_m) and those that don't (AX_n). If we start with the latter category we can show quite readily that 'the most highly symmetric' geometry is the one that is most stable. Let's use the simplest example, that of BeH_2. Figure 17.2 shows a diagram that plots orbital energy against distortion coordinate, in this case the bending of the molecule. Such diagrams are called Walsh diagrams[2] after the spectroscopist A. D. Walsh who used them just as we are doing to study the geometric preferences of molecules in their ground and excited states. (Like many other techniques of molecular orbital theory they have their origin with Robert Mulliken.) The lowest energy orbital (σ_g^+), the in-phase combination of central atom 2s and hydrogen 1s orbitals, remains roughly constant in energy on bending. The next highest orbital (σ_u^+) is rapidly destabilized on bending as a direct result of the loss of overlap between the Be 2p and hydrogen 1s orbitals. Thus the electronic energy is lowest when the molecule is linear. This is the observed equilibrium geometry of the molecule.

As a general result you can show by using similar ideas for the trigonal AH_3, tetrahedral AH_4, trigonal bipyramidal AH_5 and octahedral AH_6 molecules, that these symmetric structures are most stable for 3–6 electron pairs. In orbital terms this result is just a property of the angular form of the wavefunctions. You can simply see this by[3] using the Angular Overlap Model of Chapter 7.

> Since the ligands may well be charged due to ligand–central atom charge transfer, these symmetrical structures are those we would find too by minimizing the electrostatic energy. If I minimize the total energy of a set of charged particles on a sphere I get this self-same result. The same is true of steric interactions between the ligands. I get similar results if the inter-ligand interaction goes as $1/r^n$ in general, although whether the square pyramid or trigonal bipyramid is found for AX_5 does depend on the value of n.

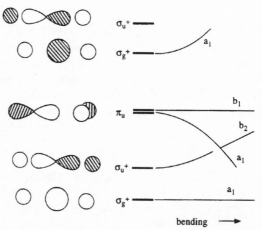

Figure 17.2
A Walsh diagram for the bending of BeH₂

This is quite true. Several different, conceptually simple, models give the same result here. It makes it difficult to pinpoint what is actually going on.

When there are more electron pairs than ligands then one or more lone pairs are generated. If the ligands are more electronegative than the central atom (the usual case), the lone pair is generated on the central atom. (Strictly speaking this will be an orbital that has more central atom character than ligand character.) Thus the structure will be distorted away from this symmetrical geometry. But what structure will the molecule have? The model that I like to use is that due to Lawrence Bartell[4]. The symmetric structure found for the AX_n case is the one where the occupied orbitals are maximally stabilized and, as a result, one where the unoccupied orbitals lie to high energy. Thus, when exploring the structure of the AX_nE system with one more pair of electrons than AX_n, we need to find the geometry which best stabilizes that high-lying orbital of the symmetric geometry which is now occupied by this extra electron pair (Structure **17.1**). The route we can usefully take employs the second-order Jahn–Teller effect.

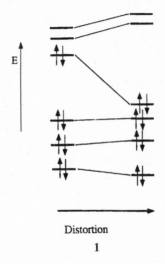

Distortion

1

The second-order Jahn–Teller effect relies, if you remember from Chapter 14, on truncating the second-order perturbation theory sum of equation (14.4) to include only the first, and hopefully the most important term, associated with the lowest-lying electronic state obtained by promotion of an electron from HOMO to LUMO. It relied on the fact that the denominator of the perturbation expression is smallest for this excitation. In fact, as we showed, the whole expression can be simplified so that instead of worrying about electronic states, all we need for operation of the scheme is knowledge of the symmetry species of HOMO and LUMO.

Yes, if there is a low-lying LUMO of symmetry Γ_j and the HOMO is of symmetry Γ_i, then the second-order Jahn–Teller active distortion coordinate is of symmetry $\Gamma_q = \Gamma_i \otimes \Gamma_j$. It will not always correspond to an allowed motion of the molecule though. In this case I expect that the geometry will remain unchanged. However, if Γ_q does correspond to a normal mode, particularly a bending mode that will change the geometry, then if the driving force is large enough, the parent structure will distort.

Precisely. Let's see how this concept can be applied to the geometry problem.

Figure 17.3(a) shows a molecular orbital diagram for a linear AH_2 molecule. The orbital occupation shown corresponds to that of BeH_2. With two pairs of valence electrons VSEPR tells us this should indeed be a linear molecule. The symmetry of the bending mode for such a molecule is π_u. Figure 17.3(a) also shows the excitation HOMO–LUMO for this electron count. It couples the HOMO (σ_u^+) with the LUMO (π_u). The second-order Jahn–Teller mode is thus of $\sigma_u^+ \otimes \pi_u = \pi_g$ symmetry. Since this does not correspond to the symmetry species of the bending mode of the molecule (π_u), the molecule remains linear. The situation is different for the molecule with one more pair of electrons (Figure 17.3(b)). Now the HOMO (π_u) and LUMO (σ_g^+) lead to a second-order Jahn–Teller mode of $\pi_u \otimes \sigma_g^+ = \pi_u$ symmetry. This does correspond to the bending motion of the molecule and thus the molecule should bend with the result that its equilibrium geometry should be non-linear. Indeed singlet CH_2 is non-linear. The stabilization of the HOMO which results (Structure **17.1**) comes about via mixing of the LUMO into the HOMO. This is shown in Figure 17.4 and shows how a lone pair develops exactly in the place where VSEPR tells us it should.

Let's have a look at the molecule (OH_2) with one more pair of electrons (Figure 17.3(c)). The symmetry of HOMO and LUMO are exactly the same as that for the system with three electron pairs, and thus the system distorts in the same way. Figure 17.5(a) shows the orbital description of the occupied orbitals. Notice that the second lone pair has not developed in the region expected from VSEPR. However, as we showed in Chapter 3, we can localize these delocalized electrons (Figure 17.5(b)), just as in CH_2, to produce two lone pairs that are the characteristic 'rabbit ears' of water. The same process may be performed on the two bonding orbitals, just as we did for BeH_2.

Thus the four localized orbitals of water which result, can be made to point toward the vertices of a tetrahedron (Structure **17.2**) just like those of methane.

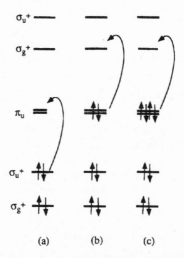

Figure 17.3

(a) A molecular orbital diagram for a linear AH_2 molecule. It also shows the HOMO–LUMO excitation for the electron count (two valence pairs) appropriate for BeH_2; (b) the HOMO–LUMO excitation for the electron count (three valence pairs) appropriate for singlet CH_2; (c) the HOMO–LUMO excitation for the electron count (four valence pairs) appropriate for water

2

Yes, for the AX_2E molecule with only one lone pair the localized and de-localized models place it in the same region of space, but for more than one lone pair (AX_2E_2) then some rearrangement of the de-localized picture is required. Remember, however, from Chapter 3 that these are quite arbitrary constructs. The total electron density is independent of the choice of wavefunction within the restrictions we described.

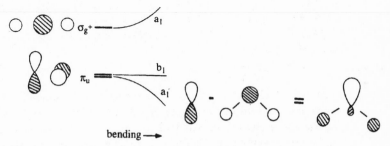

Figure 17.4
The stabilization of the a_1 HOMO (see Structure 17.1) for singlet CH_2 as a result of mixing of HOMO and LUMO. A lone pair develops where VSEPR expects it to be

The description of the distortion of the planar AH_3 molecule proceeds along the same route. The out-of-plane bending mode is of a_2'' symmetry and the in-plane bend of e' symmetry. For BH_3 in Figure 17.6(a), the HOMO (e') and LUMO (a_2'') lead to a second-order Jahn–Teller mode of $e' \otimes a_2'' = e''$ symmetry. Thus the molecule is stable as a trigonal plane. There is no vibrational mode (and no bending mode) of this symmetry. For NH_3, the HOMO (a_2'') and LUMO (a_1') lead to a second-order Jahn–Teller mode of $a_2'' \otimes a_1' = a_2''$ symmetry (Figure 17.6(b)). This is the symmetry of the out-of-plane bend. The molecule thus bends from the flat D_{3h} structure to a C_{3v} pyramid. As before, HOMO

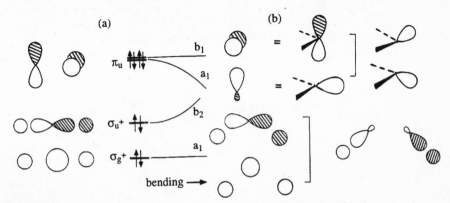

Figure 17.5
(a) A picture similar to that of Figure 17.4 that also shows (b) how the occupied orbitals may be localized to give bonding and non-bonding orbitals for water

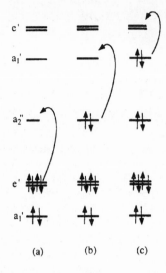

Figure 17.6
(a) A molecular orbital diagram for a trigonal planar AH_3 molecule. It also shows the HOMO–LUMO excitation for the electron count (three valence pairs) appropriate for BH_3; (b) the HOMO–LUMO excitation for the electron count (four valence pairs) appropriate for NH_3; (c) the HOMO–LUMO excitation for the electron count (five valence pairs) appropriate for FH_3

and LUMO mix to give a lone pair that points towards the fourth vertex of the tetrahedron (Figure 14.4). With one more pair of electrons (Figure 17.6(c) for ClH_3 (ClF_3 is the known species)) the HOMO (a_1') and LUMO (e') lead to a second-order Jahn–Teller mode of $a_1' \otimes e' = e'$ symmetry. The molecule bends in-plane towards a T-shape. (There is also incidentally a bond-stretching vibration of e' symmetry that makes the C–H bonds inequivalent too.) Just as for the AH_2 molecule, the lone pairs from the delocalized model need to be localized to match those of the VSEPR scheme. Other molecules can be viewed in the same way.

Similar considerations apply to other stoichiometries and this second-order Jahn–Teller orbital model is a useful one for understanding the geometries, not only of these AH_n molecules, but, as shown in Chapter 14, those of other systems.

One interesting molecule we should discuss here is that of XeF_6. VSEPR counts seven valence electron pairs but the molecule is only dynamically distorted away from the octahedral geometry expected for six pairs. By this we mean that the molecule is executing a soft geometrical motion around a parent higher symmetry structure[5]. The second-order Jahn–Teller approach allows us to understand this. Figure 17.7(a) shows an orbital picture for octahedral XeF_6. Notice that the HOMO is of a_{1g} symmetry and that there is a low-lying LUMO (t_{1u}). The second-order Jahn–Teller active mode is of $a_{1g} \otimes t_{1u} = t_{1u}$ symmetry. As a result XeF_6 experiences a distortion away from octahedral along a t_{1u}

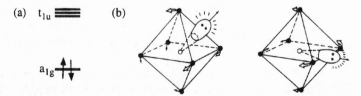

Figure 17.7
(a) The HOMO and LUMO for octahedral XeF_6; (b) the geometry that results from a soft t_{1u} bending mode (adapted from reference 5)

coordinate, one that leads to the generation of a lone pair (Figure 17.7(b)) either along an edge or in a face of the octahedron.

However, for the heavy elements in this part of the Periodic Table, the $6s^2$ pair of electrons is often stereochemically inert as a result of a general relativistic contraction[6] of s orbitals relative to p. This means that, although the second-order Jahn–Teller denominator may be small, the numerator, which involves overlap with the central atom 6s orbital may be small too. So we should expect a rather soft distortion away from the octahedral structure and a dynamic second-order Jahn–Teller effect, as is observed.

Yes, some formally seven pair molecules (such as AX_6^{2-}; A = Se, Te; X = Cl, Br) are found as regular octahedral structures where there is no distortion. Others are distorted, such as SbX_6^{3-}. $TeCl_4(thiourea)_2$ is known in two forms, one distorted, the other not[7].

So here is one view of the molecular orbital underpinnings of the VSEPR scheme. While we have to work harder to generate the geometry of the ground state on the orbital scheme, one of the advantages of the orbital picture is that it is much easier to view other properties of the molecule, those of the excited state for example. There is also a satisfaction in being able to use the same orbital model throughout all of chemistry.

But where is the connection to the VSEPR scheme through the 'Pauli repulsion forces'?

We have already noted that the geometries of these molecules are well reproduced by orbital considerations that do not formally include electron–electron interactions. We have also shown in Figure 17.5 that one can generate VSEPR pairs from a delocalized picture, and recall from Chapter 3 that the total energy is the same using either descrption. So we must search inside the one-electron orbital picture for the connection with the Pauli Principle if there is one. Figure 17.2 gives us some clues. The 'repulsion' between the two pairs of electrons in BeH_2 that leads to a linear rather than bent geometry is reflected by the destabilization of the σ_u ($1b_1$) orbital (occupied by one pair of electrons) on bending. The interaction among the three pairs of electrons in singlet CH_2 leads to a bent molecule. Its geometry is set by the balance of occupied orbitals going up and down in energy on bending. Now the Pauli Principle tells us that only two electrons, one of each spin, may occupy a given atomic or molecular orbital. Thus the electron configuration of BeH_2 is $\sigma_g^2\sigma_u^2$ and of singlet CH_2, $1a_1^2 1b_1^2 2a_1^2$. Without the restrictions of the Pauli Principle we could imagine a configuration for CH_2 of $1a_1$. A molecule with this configuration would not have a very strong geometric preference since the $1a_1$ orbital does not change in energy very much on bending but would be one where the construction of localized orbitals would be inapplicable.

So it is the antisymmetrization of the total wavefunction (Chapter 3) demanded by Pauli, which forces pairs of electrons into different orthogonal orbitals whose energies are determined by the geometry. It is this that leads to geometric differences with electron count.

Yes, the 'repulsion' between the localized pairs of electrons comes from this source. We noted a similar effect in Chapter 15, where by forcing two electrons into a bonding orbital and two into an antibonding orbital the overall effect is a destabilizing one (via a similar Pauli repulsion). Here the origin of the destabilization comes from a different source—the geometry change.

References

1. Gillespie, R. J. and Hargittai, I., *The VSEPR Model of Molecular Geometry*, Allyn and Bacon (1991).
2. Walsh, A. D., *J. Chem. Soc.*, 2260, 2266, 2288, 2296 (1953).
3. Albright, T. A. and Burdett, J. K., *Problems in Molecular Orbital Theory*, Oxford University Press (1992), p 104.
4. Bartell, L. S., *J. Chem. Educ.*, **45**, 754 (1968).
5. Bartell, L. S. and Gavin, R. M., *J. Chem. Phys.*, **48**, 2470 (1968).
6. Pyykkö, P., *Chem. Rev.*, **88**, 563 (1988).
7. Husebye, S. and George, J. W., *Inorg Chem.*, **8**, 314 (1969); George, J. W., Husebye, S. and Mikalsen, Ø., *Acta Chem. Scand.*, **29**, 141 (1975).

General References and Further Reading

Bader, R. F. W., *Atoms in Molecules*, Oxford University Press (1994).
Edmiston, C., Bartleson, J. and Jarvie, J., *J. Amer. Chem. Soc.*, **108**, 3593 (1986).
Gillespie, R. J., *Molecular Geometry*, Van Nostrand-Rheinhold (1972).
Gillespie, R. J. and Robinson, E. A., *Angew. Chem. Int. Ed.*, **35**, 495 (1996).
Gimarc, B. M., *Molecular Structure and Bonding*, Academic Press (1979).
Fergusson, J. F., *Stereochemistry and Bonding in Inorganic Chemistry*, Prentice-Hall (1974).
Sidgwick, N. V. and Powell, H. M., *Proc. Roy. Soc.*, **A176**, 153 (1940).

Epilogue

Well, by the end of our dialog we have covered quite a bit of ground, although there are several interesting areas we didn't even touch upon of course. There is the comment of the Young Scientist that I mentioned in the Prologue but it is appropriate too at this point concerning the way we have studied the topics in this book. 'It is not so much a question of difficulty or time, as of being prepared to think. The subject is not one to read in spare time in railway carriages, or while lunching in a restaurant.'

That *is* a good one. You mentioned too an apt comment of Charles Coulson's in the Prologue that is also useful to recall here. 'The rôle of quantum chemistry is to *understand* the elementary concepts of chemistry ...' (my italics). The question of understanding is indeed the important one, irrespective of whether the result comes from an experiment, a simple theoretical model or from a complex calculation. And this, like many other endeavors, is not a trivial task.

But just like Hume-Rothery's Young Scientist we must not be complacent. Most of the questions we have tackled have been, like many of his, ones that we understand, some in depth, some not quite so thoroughly. But he looked to the future as well. One of the great challenges to chemical science today is to provide answers to questions that we haven't even discussed. For example, consider the stability of compounds. They can be made, their structures increasingly easy to determine and increasingly easy to understand using the models used in this book. But why can they be made in the first place? How do we rationalize their structures, reactivity and stability with a global picture, rather than rather narrow and specific ones tied to given collections of atoms? We have not mentioned, either, the great progress in recent years in the study of biological molecules, held together by the weak forces we have barely mentioned in our discussions. The level of our understanding concerning their structure is a little like that of the Old Metallurgist and his work in the foundry.

Certainly this is a sobering thought. We have indeed looked to the past and present in studying existing models but I'm sure that the future will be just as exciting and as unpredictable. Let's conclude though on a different note. At the end of Hume-Rothery's book the Metallurgist (now retired) and the Scientist (now middle-aged) drink a toast to their publishers and friends. Well, we have come to the end of our discussion. Who should we thank for their help?

We can describe a major part of this in terms of a scurrilous corruption of the first-order correction to the ground-state wavefunction, $|0\rangle$, using perturbation theory. $|0'\rangle$ is the new wavefunction (our present state of understanding) arising through the perturbation $\mathcal{H}'$ (which I think we would like to assign to the thought process).

$$|0'\rangle = |0\rangle + \sum_j{}' |\langle 0|\mathcal{H}'|j\rangle|/(E_j - E_0)|j\rangle$$

Here $j =$ Mulliken, Pauling, Lennard-Jones, Coulson and Hoffmann to mention just five. For these the coefficient $|\langle 0|\mathcal{H}'|j\rangle|/(E_j - E_0)$ is particularly large and their ideas have influenced us enormously. However, on moving to the immediate present, Tom Albright read the manuscript and made several useful comments about the dialog as did a number of anonymous reviewers. We owe them our thanks for their insights. Graham Fleming persisted in asking questions about what is behind the chemical bond. There are also several generations of students who have played your rôle in asking probing questions about these topics at The University of Chicago.

Index